Cognitive Space and Patterns of Deceit in La Fontaine's *Contes*

EMF CRITIQUES

Cognitive Space and Patterns of Deceit
in La Fontaine's Contes
COPY EDITOR
Marie Hertzler
PRODUCTION EDITOR
Cara Welch

Cover Design by Dallas Pasco

Catherine M. Grisé

COGNITIVE SPACE AND PATTERNS OF DECEIT IN LA FONTAINE'S *CONTES*

ROOKWOOD PRESS
Charlottesville

520 Rookwood Place
Charlottesville, Virginia 22903-4734, USA

Printed in the United States of America.
02 01 00 99 98 5 4 3 2 1

This book is printed on acid-free paper.

Library of Congress Cataloging-in-Publication Data

Grisé, Catherine M.
Cognitive space and patterns of deceit in La Fontaine's Contes / Catherine M. Grisé.
p. cm. -- (EMF critiques)
Includes bibliographical references and index.
ISBN 1-886365-09-1 (lib. bdg.)
1. La Fontaine, Jean de, 1621-1695. Contes et nouvelles en vers. 2. Semiotics and literature. 3. Deception in literature. I. Title. II. Series.
PQ1809.G75 1998
841'.4--dc21 97-50295
CIP

The EMF Critiques series is sponsored by the Department of French, University of Virginia

TABLE OF CONTENTS

This book has been published with the assistance of a grant from the French Department of the University of Toronto and from St. Michael's College, University of Toronto.

COGNITIVE SPACE AND PATTERNS OF DECEIT IN LA FONTAINE'S *CONTES*

INTRODUCTION

The *Contes* of La Fontaine, immensely popular with readers and frequently imitated by other poets up to the end of the nineteenth century, have not fared well with literary critics until recently. Even in the seventeenth century Charpentier commented on the dichotomy between the critic Furetière's censorious attitude and the enthusiastic response of the public: "Le plus impie et le plus sale de tous les hommes [Furetière] reproche à La Fontaine quelques contes que le public excuse aisément par la manière agréable et ingénieuse dont le poète les a tournés" (Mongrédien 147).

La Fontaine's *Contes* belong to the vast corpus of licentious literature whose roots extend back into the oral culture of ancient folklore. This tradition was mediated to La Fontaine chiefly through Ariosto, Boccaccio, and the French *nouvellistes* of the sixteenth century. It emerges from his pen, however, as a work steeped in the ethos of the seventeenth-century and filtered through the literary consciousness of one of the most talented of French poets. And yet, the scabrous subject matter has been a major stumbling block to the acceptance of the *Contes* into the canon of French literature. Until not so long ago, the prevailing approach was a paradoxical silence: "En parler pour les blâmer," commented the nineteenth-century critic, Nisard, "serait pruderie; les louer, ils n'en ont pas besoin. Ces livres-là ne font que trop leur fortune d'eux-mêmes; n'en rien dire est le plus sage" (163).

The characteristic licentiousness of the subject matter coupled with a deliberately "irregular" style accounts for the genre being considered not only minor, but definitely a *genre bas* and unfit for critical commentary. The *Contes*, unlike the *Fables*, were not "presentable" to the young Dauphin. Nevertheless, they have enjoyed great popularity; over a hundred editions were published in the nineteenth century alone.[1] Despite the popular success of the *Contes*, historians and critics of classical French literature have, until recently, judged them to be unsuitable for scholarly perusal.

COGNITIVE SPACE AND PATTERNS OF DECEIT

The first of La Fontaine's *Nouvelles en vers* were published late in 1664. Despite the controversies and scandal they provoked, La Fontaine continued to compose and publish his tales (1665, 1666, 1667, 1669, 1671, 1674, 1685) until a few years before his death in 1695.[2] La Fontaine restricted his tales almost exclusively to accounts of sexual deception. The narratives he chose to retell present typical trickster motifs in sexual contexts.[3] La Fontaine's choice of this particular focus is one of the strong identifying characteristics of the *Contes*. The licentious subject matter is an essential element of the genre and provides a playful context for the narration, much as the animal world does in the *Fables*.

The rehabilitation of the *Contes* as a subject of scholarly investigation was initiated by John Lapp in 1971 with the publication of *The Esthetics of Negligence: La Fontaine's* Contes. Since then interest in the *Contes* continues to grow, in the genre itself (John D. Lyons, "D'une vérité sans effet: La Fontaine et la théorie de la nouvelle" and Jürgen Grimm, "'Le Faucon.' La Fontaine et Boccace: Proximité et distance") along with its themes (Jean-Pierre Collinet, "L'Amour et son expression poétique dans les *Contes* de La Fontaine"), its parody of casuistry (Martha Houle, "La Fontaine's Fun with Casuistry in 'Le Cas de Conscience'"), and its irony (Donna Kuizenga, "La Fontaine's 'Le Faucon': A Lesson of Experience."). A number of studies analyze the narrative structures, using a semiotic approach (Jane Merino-Morais, *Différence et répétition dans les* Contes *de La Fontaine*) or a poetics of narrative; the latter approach has led to studies of focalization (Madeleine Defrenne, "La Fontaine et les jeux de la focalisation"), of the cognitive dimension (my article, "Erotic Dimensions of Space in La Fontaine's 'La Fiancée du roi de Garbe'"), and of figuration in the text (Michael Vincent, "Reading (through) the Veil: 'Le Tableau'" and Anne L. Birberick, "From World to Text: The Figure of the Nun in La Fontaine's *Contes*").[4]

THE COGNITIVE DIMENSION AND FICTIONAL CHARACTERS

My study of cognitive space and patterns of deceit in La Fontaine's *Contes* is grounded in a theory of the cognitive dimension in literature and of frame analysis. Since the 1950s literary theorists have been investigating the degree of knowledge accorded to the narrator and to the narratee in fiction. Terms such as the "omniscient narrator" and the "implied reader" have become commonplaces.[5] Literary criticism has, however, explored to a much lesser extent the functioning of the characters' knowledge within the fictional world.[6]

How the characters within the story world (or the "diegetic world" according to Genette) share knowledge or conceal it from each other shapes the various configurations of the fictional narrative, from unexpected turns of plot to the interaction of characters. This cognitive dimension of the text is the locus of the circulation of knowledge and as such directly influences the reader's experience of the text, its ironies, its suspense, its revelations.

A.-J. Greimas proposed the theoretical term "cognitive dimension" to describe that aspect of the text to be explored through an analysis of "knowledge about events" in the narrative utterance.[7] Semioticians and others who apply epistemological theories to the study of literature, linguistics, film, and painting are methodically exploring the cognitive dimension.[8] Insofar as the literary text is concerned, however, the primary focus of interest is still what semioticians call the level of the *énonciation*; Jacques Fontanille, for example, has astutely analyzed the presence of an observer in the text, variously defined as *focalisateur*, *spectateur*, *assistant*, or *assistant participant*.[9]

The character in fiction is a problematic area of critical debate. After the Romantic and post-Romantic simplistic equation of characters with real people, structuralists, linguists, semioticians, and New Critics reacted excessively by minimizing the referential qualities of literary characters.[10] In 1979 Greimas and Courtés declared *personnage* an outdated term in the *Dictionnaire raisonné*'s succinct entry:

> Employé, entre autres, en littérature et réservé aux personnes humaines, le terme de personnage a été progressivement remplacé par les deux concepts—plus rigoureusement définis en sémiotique—d'actant et d'acteur. (274)

Nevertheless, the tenacity of the concept of the fictive character in literary criticism was confirmed only a few years later at a colloquium on the *personnage*:

> La question du personnage s'est posée à nous comme nécessité d'éclaircir une situation ambiguë: après les excès auxquels s'était livrée une critique qui ne faisait aucune différence entre personnage et personne, la réaction avait été saine et violente. Ce fut la mort du personnage, perpétrée par la critique et consommée par le Nouveau Roman français. Pourtant le personnage, expulsé par la porte, revenait par la fenêtre et refusait de se laisser réduire à l'état d'actant-acteur-fonction. (Ezquerro 108)

The concept of fictional characters is being rehabilitated through the works of Baruch Hochman, Shlomith Rimmon-Kenan, and, especially, Seymour Chatman, who as early as 1978 explored in *Story and Discourse* how narratives generate in the reader's consciousness a spatial world inhabited by characters. Nonetheless, finding a middle ground between semiotic reductiveness and a mimetic equation of characters with real people remains a challenge for literary theory.[11]

The cognitive dimension is then, in the fictive world, the locus of representation of the characters' knowledge and beliefs. Knowledge states, or mental states, are part of the composition of a character, and constitute constantly changing attributes, unlike other more stable attributes of age, sex, or personality traits.[12] Within the cognitive space of the text the circulation of knowledge determines to a large extent both the subjective transformations of the characters and the narrative action in the story world. I explore this diegetic world in the *Contes* of La Fontaine, a textual space inhabited by characters whose manipulation of knowledge through various covert strategies inscribes in the text patterns of deceit.

DECEIT AS A COGNITIVE MANIPULATION

Deceit in its diverse manifestations is a subject of intense research on the part of philosophers, sociologists, psychologists, and biologists. An analysis, such as I propose, of deceit in the fictional world, benefits from the hypotheses and conclusions of those who pursue theories about deception in the "real" world. Deception takes on many forms in the human and nonhuman worlds. Biologists document cases of camouflage, mimicry, and delusion. Chameleons change color to blend in with their surroundings and hide from predators. Fireflies mimic one another in using light signals for sexual communication; some females of one species are said to send false signals to a male of another species, then lie in wait to devour the unsuspecting male. There is even evidence of apparently intentional animal deception. A chimp was observed pretending not to know where food was hidden, but later, when the other chimps were asleep, it went directly to the hiding spot to eat the food. Apes trained to use language are reported to have lied to their trainers.[13]

Biologists debate, however, about whether even the most intelligent of animals, the nonhuman primates, can attribute beliefs, knowledge, or motives to others; if they cannot, intentional deception would be an impossibility. The problem of deceit in the animal world is whether chimpanzees, for example, can create a desired state of mind in another and recognize that their own behavior can influence that belief. Attribution of knowledge to another "demands some ability to represent simultaneously two different states of mind. To do this an individual must recognize that he has knowledge, that others have knowledge, and that there can be a discrepancy between his own knowledge and theirs" (Cheeney and

Seyfarth 207). Human-like deception requires that a signaler create or support a false belief in another. One chimpanzee may, in fact, be unable to recognize that another chimpanzee can possess a belief incompatible with his (or her) own and therefore, "may be unable to accommodate simultaneously both a true and a false belief" (Cheeney and Seyfarth 218).

Deception is a human universal that presents basically the same features whether it occurs in a seventeenth-century context or in our own period of time. It is classically defined as "an act that is *intended* to foster in *another person* a belief or understanding which the deceiver *considers* false" (Zuckerman et al. 3). Two of the most contentious problems that are debated about deception are whether intentionality is essential to an act of deception and whether self-deception is possible. In my analysis I consider that the classical definition needs to include forms of self-deception and some of those less intentional acts of deception where human conduct seems programmed unconsciously, much like the chameleon or the firefly.

Deceit is a cognitive event that depends on the fabrication and manipulation of knowledge. The fundamental cognitive transformations set into play by deception are easily described. First of all, an act, such as lying, changes the belief of the target of the deceptive action. The deceivee gives credence to the deceiver when a fiduciary relationship—a relationship of faith or trust—has already been established. When this relationship is lacking, however, the process of deception is more complex and requires further confirmation of the grounds of belief. If the verification procedure succeeds, the fiduciary contract is established, and acceptance of the false belief ensues. The purpose of the deceit is achieved when the deceivee acts in accordance with the false belief. The deceivee may, in the end, discover the deceit, and, in some cases, punish the deceiver. In the act of deception, basic structures of exchange and substitution underlie the various strategies: falsehood is substituted for truth; exchange pacts are negotiated under false pretenses; and contracts are subverted by covert substitutions.

FRAME ANALYSIS AND PATTERNS OF DECEIT

I situate deception within the cognitive dimension and account for the circulation, withholding, and fabrication of knowledge within the story world by focusing on the characters' epistemic frames. Seymour Chatman's statement approaches my view of the inner diegetic space of fiction as a locus of cognitive transformations:

> Only characters reside in the constructed story world, so only they can be said to "see," that is to have a diegetic conscious-

> ness that literally perceives and thinks about things from a position within that world. Only their "perspective" is immanent to that world. (*Coming to Terms* 146)[14]

I am not proposing a study of "focalization" nor of "point of view" as an approach to the analysis of the cognitive dimension in which the characters of La Fontaine's *Contes* function.[15] Although focalization and point of view are popular and useful critical terms, they are no more appropriate for analyzing deceit in a literary text than for describing dramatic irony.[16] We need a theory that takes into account simultaneously the circulation of knowledge within the story world and its perception by the extradiegetic agents—the narrator, the narratee, and the reader.[17] Dramatic irony, like deceit, involves comparing discrepant knowledge states. The audience, which incorporates into its own cognitive frame all the knowledge that has been communicated about the individual characters' knowledge, compares divergent mental spaces. Take two simple examples, one from Molière's *Tartuffe* and the other from Shakespeare's *Romeo and Juliet.* When Orgon hides under the table in *Tartuffe* the audience knows he is there, and so does Elmire. Only Tartuffe has a false framing of the situation. All of Tartuffe's words, as they are being pronounced, are filtered through the audience's interpretive frame, "Tartuffe is under the table." When Romeo returns to find Juliet in the tomb, he falsely believes she is dead. The audience has within its true knowledge frame the information that she is not dead. Again, all his actions and words have as their referent the true belief state of the audience. Simultaneously, the audience sees two frames that are contradictory, the false frame of Romeo or of Tartuffe and its own "true" framing—shared by Elmire and, to some extent, Orgon, in *Tartuffe*. Conventional theories of focalization and point of view fail to adequately account for the presence of contradictory mental states.[18]

The ability to hold simultaneously contradictory mental states may prevent most animals from being capable of deception: they cannot recognize their own mental state and they cannot attribute beliefs to others. The authors of *How Monkeys See the World* imagine an audience of chimpanzees watching *Romeo and Juliet*:

> Unlike monkeys, chimpanzees seem to understand each other's goals and motives, but may be unable to attribute false belief to others (like small children), do not recognize the discrepancy between their own states of mind and the states of minds of others. How would Romeo's death appear to an audience of monkeys, unable to distinguish between their own beliefs and Romeo's? (Cheeney and Seyfarth 254)

The approach of frame analysis is more useful for my purposes than the concept of focalization or the semiotic schematizations of manipulation and deceit.[19] The term "frame" is most often associated with anthropology, sociology, and linguistics.[20] Scholars such as Bateson, Hymes, Goffman, Frake, and Fillmore adopt an interactive model for their theories, considering the framing activity as an interpretive process. If, for example, a television announcer is commenting on an imminent nuclear disaster, the frame of interpretation needs clarification for the viewer: Is this truly a newscast or is it part of a science-fiction film?

Scholars of cognitive science and information theory discuss the construction of mental spaces in the process of passing from one belief state to another.[21] They use the term "mental space" as a metaphorical way of designating subjective constructions of knowledge, or the processing of information by an individual. Just as the world of dreams seems to occupy another space, so do beliefs, hopes, or fears. The spaces are, we might say, framed, or contextualized, by a myriad of factors that act as guides for the interpretation of experience or information.[22] These factors build cognitive space inasmuch as they generate interpretive frames. The space encompassed by the frames is mental space, not topographical space.

Frame analysis examines how human experiences are organized into understandable contexts. Events are framed or set into perspectives that aid in their interpretation. For example, a violent scene on the stage is perceived differently from the same scene viewed in a subway station. Erving Goffman explains:

> I start with the fact that from an individual's particular point of view, while one thing may momentarily appear to be what is really going on, in fact what is actually happening is plainly a joke, or a dream, or an accident, or a mistake, or a misunderstanding, or a deception, or a theatrical performance, and so forth. (10)

When the frame of interpretation shifts suddenly, the framing activity and the mode of framing become most apparent. Television provided an example of a shifting frame of interpretation some years ago on April Fool's Day when at the end of the news a lengthy documentary explained the harvesting of spaghetti from spaghetti trees in Italy. At the outset viewers understood the information in the factual framing mode set up by a news broadcast. Then, as the absurdities of the reportage accumulated, the viewer inevitably switched to a counterfactual mode of interpretation in which the full irony of the situation could be appreciated.

THE METHODOLOGY OF FRAME ANALYSIS

My theory of frame analysis represents deceit patterns in the fictional world; the metaphor of a "frame" of knowledge effectively conceptualizes and expresses in a diagram the states of knowledge of the characters and the extradiegetic agents (the narrator, the narratee, the reader). I consider the epistemic constructs in the text as bound by frames, not static frames, but interactive interpretive frames, much as Gregory Bateson and Erving Goffman describe them. The diagrams in which I illustrate more graphically the patterns of deceit are, like all schematizations, oversimplifications. They are valuable, however, because they readily translate the various levels of opacity between characters and take into account simultaneous discrepancies in the interpretation of situations. While Erving Goffman did not devise visual diagrams, his use of the term "frame" invites this type of graphic elaboration:

> Observe that for those in on a deception, what is going on is fabrication; for those contained, what is going on is *what* is being fabricated. The rim of the frame is a construction, but only the fabricators see it. (84)

Deceit is a manipulative process that involves shifting the deceivee's interpretation of information into either a counterfactual frame, a doubtful frame, or a hypothetical frame.[23] The shifting of perspectives, or frames of interpretation, is accomplished by what I term space-builders, such as lying, false promises, or other forms of cognitive manipulation. When a deceiver's lie falsifies the deceivee's belief space, the interpretive frame of the victim is altered. The belief space generated is false—a counterfactual representation of reality—and contrasts with the true belief space of the agent of deceit. When doubt and suspicion surround a character's perception of a situation, the doubtful frame acts as a filter through which a scene is interpreted. When a deceiver's false promise produces another kind of false representation—a distorted view of the future—I call this frame hypothetical because it is grounded on a false hypothesis. In any of these situations both the deceiver and a knowledgeable spectator simultaneously perceive two mental spaces, the true and the false, and comprehend the contradiction between the two.

An example of counterfactual framing occurs in La Fontaine's tale, "Le Muletier,"[24] in which the deceit is accomplished by the space-builder of disguise. A mule driver appears at the queen's bedroom door disguised as the king; he gains access to her bed without her understanding that she is committing a sexual transgression. This pattern of deceit can be represented by a diagram in which a broken line represents the queen's false frame of belief, while the deceiver's factual framing, indicated by an unbroken line, comprehends the whole situation.

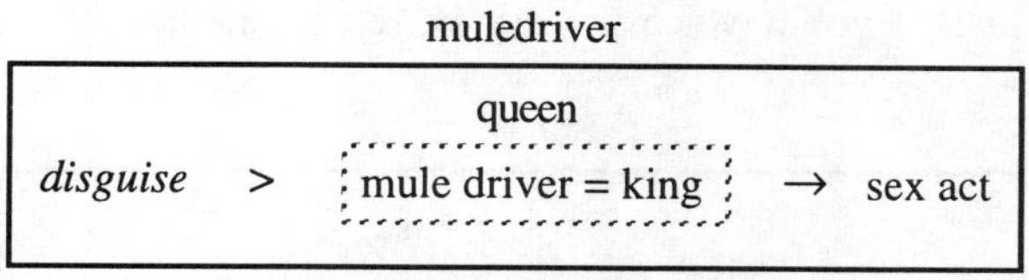

> creates a cognitive space
→ resultant action
Figure I.1

Doubtful framing takes place when doubt or suspicion make the perception uncertain. In this case the person who is being deceived tentatively believes the information provided by the deceiver. The fiduciary relationship is not yet established; further verification is required. A doubtful mental space is relatively weak because the subject is aware of the uncertain nature of the basis for belief. When further evidence reinforces and confirms the initial information, the doubtful frame shifts to a false or counterfactual interpretation of the data. Doubtful framing—indicated below through a frame of question marks—is present at the outset the second woman's tale in "La Gageure des trois commères" : the husband doubts his wife and her lover's protestations that they were not making love beneath the pear tree.

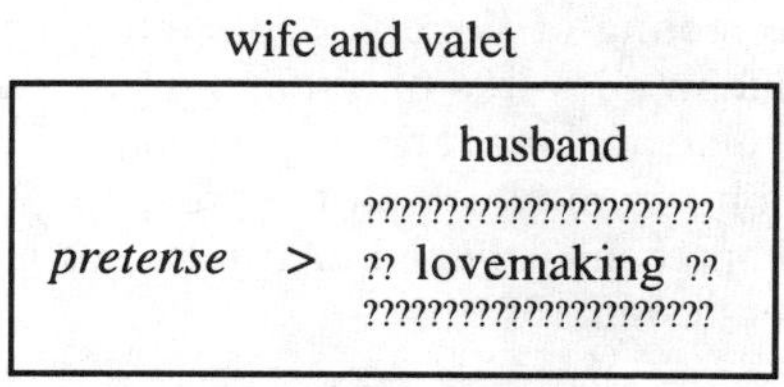

Figure I.2

Hypothetical framing functions when the cognitive space-builder is a false promise. A false anticipatory space is generated when the deceiver promises in a hypothetical fashion the attainment of a desirable future goal provided that the deceivee consent to the conditions of an agreement. The seduction of Lise in "Comment l'esprit vient aux filles" is accomplished by manipulating her into falsely believing in the attainment of a hypothetical future state of intelligence—only to be achieved, however, if she consents to sex with Bonaventure, the priest. In the following diagram we see within Bonaventure's all-encompassing mental space Lise's false framing of the situation, indicated by a frame of broken lines. She has constructed a hypothetical cognitive space, indicated by a frame of exclamation marks, within which she sees herself as intelligent.

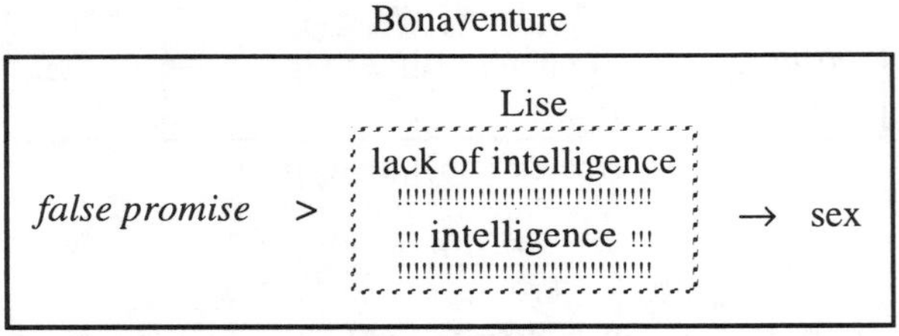

Figure I.3

COGNITIVE CLOSURE

Barbara Herrnstein Smith demonstrates in her book on poetic closure a variety of devices ranging from thematic allusions to typographical features that come into play at the end of a poem:

> Closure occurs when the concluding portion of a poem creates in the reader a sense of appropriate cessation. It announces and justifies the absence of further development; it reinforces the feeling of finality, completion, and composure which we value in all works of art; and it gives ultimate unity and coherence to the reader's experience of the poem by providing a point from which all the preceding elements may be viewed comprehensively and their relations grasped as part of a significant design. (36)[25]

While all the usual poetic devices operate to bring about closure in the *Contes*, the patterns of deceit that are so prominent in the cognitive dimension require special attention to the way closure is effected on this level of the text. Within the cognitive dimension of a Lafontainian *conte* closure is effected by the narrator's final gaze at the partitioning of the cognitive space among the participants in the deceit process. In some tales the deceivee remains blissfully ignorant at the end; in others, discovery of the dupery brings a final sense of retribution. In most instances my diagrams include a schema of the cognitive closure.

Introduction

FRAME LAMINATIONS

Frame analysis is a method that takes into account simultaneously the circulation of knowledge within the story world and its perception by the extradiegetic agents—the narrator, the narratee, and the reader. In all my diagrams the diegetic frames are surrounded by a lamination of *observateurs* (which are are understood but not graphically represented). The outermost cognitive frame belongs to the reader. The extradiegetic first-person narrator occupies the next frame, within the text, while the implied reader, the textual narratee (who sometimes interrupts the narration), and the sporadic narratees—the narrator addresses from time to time husbands, wives, parents, French women, nuns—occupy frames progressively closer to the diegetic frames. In this study, however, the primary focus is deception and centers on the inner story world.

THE CORPUS

The corpus to be analyzed comprises all the *contes en vers* published in La Fontaine's *Contes et nouvelles en vers*. The *conte en vers* genre, which has been variously termed the *conte gaillard en vers*, the *conte libertin*, or the *conte galant*, is a specific subgenre of the *conte*.[26] This subgenre can be described as a lengthy poem presenting, from the perspective of a narrator who is not part of the tale (Genette's "narrateur extradiégétique-hétérodiégétique," 255), a licentious narrative told in a ludic mode and involving human dupery. The *conte en vers* with its specific characteristics was invented by La Fontaine; its status as a distinct genre was confirmed in the eighteenth and nineteenth centuries by his imitators, by critics, and by editors of his and other authors' *contes en vers*.

La Fontaine's prefatory comments to the *Contes* (Collinet 555-57, 603-05) describe the genre as dealing with a licentious subject, "la nature du conte le voulait ainsi," treated with playful humor, "cecy est jeu," and typified by an irregular style, "ces sortes de négligences qu'il ne se pardonnerait pas lui-même en un autre genre de poésie." He highlights the primacy of narrativity: "Ce n'est ni le vrai ni le vraisemblable qui font la beauté et la grâce de ces choses-ci; c'est seulement la manière de les conter." Well aware of having invented a new genre and referring to himself in the third person, La Fontaine states in the 1666 Preface: "Il s'est véritablement engagé dans une carrière toute nouvelle, et l'a fournie le mieux qu'il a pu."

Before the end of the seventeenth century the poet Saint-Glas acknowledges in the Preface to his *Contes nouveaux en vers* (1672) that La Fontaine created the genre: "Comme je ne suis pas le premier qui écris en ce genre, je trouve non seulemnt le chemin frayé, mais encore les difficultés ostées." Most critics and editors emphasize the erotic subject matter as the most salient feature of the *conte en vers*. The seventeenth-century editor Callières admires La Fontaine's talent for presenting crude subjects with an elegance of style, "l'adresse de l'ouvrier, lorsqu'il est obligé de raconter certaines avantures qu'il est difficile d'exprimer honnêtement, ou d'en réciter d'autres qui paroîtroient un peu trop libres, si elles n'étoient assaisonnées de certains adoucissemens propres à en corriger le venin" (154). In 1734 the critic Rémond de Saint-Mard highlights the libertinage of the *conte* in his *Réflexions sur la poésie en général*: "[A]u lieu de nous faire acroire, comme les autres genres de poësie qu'il va réformer nos moeurs, on diroit qu'il prend à tâche de nous les gâter: il peint pour cela la volupté avec les couleurs les plus vives, et quelle volupté? Ce n'est seulement pas celle du roman; c'est cette volupté des sens qui, aprêtée sans tant de façons, se nomme libertinage" (143). In 1802 J.-F. Guichard defends the licentiousness of his own *contes en vers* by commenting, "l'excuse n'est-elle pas dans le genre même?" (vi).

A small number of poems published in the *Contes* (Collinet edition) are excluded from my study because they do not properly speaking belong to the genre of the *conte en vers*. The corpus I deal with does not include the epigrams and minor poems (586-600) published at the end of the first collection with the avowed purpose of enlarging the volume—"pour le grossir"—as La Fontaine himself explained in his 1665 Preface. The epigram is closer to the *bon mot* genre and highlights a *facétie verbale* rather than narrative. In the Deuxième and Troisième Parties I do not consider the epigrams ("Le Villageois qui cherche son veau," "Le Bât," "Le Baiser rendu," "Epigramme" [661, 755-57], the imitations of Anacreon, and "Le Différend de beaux yeux et de belle bouche" (758-63). From the *Nouveaux Contes* I exclude the stances, "Janot et Catin" (872-74). "Clymène" (777-804), is not usually included among the *Contes* by La Fontaine's editors; Collinet acknowledges in a note that "elle ne se laisse pas ranger dans un genre déterminé" (1442). On the other hand, Collinet (unlike other modern editors) places "La Matrone d'Ephèse" and "Belphégor" (508-21) among the fables. I consider both to be *contes*, not fables, because they fulfill all the criteria of the *conte en vers* genre. The corpus I consider comprises, then, the 52 tales that are listed in Appendix A.[27]

***

The narrative poetics of the *Contes* includes as an essential component an epistemological focus. The following chapters categorize the *Contes* on the basis of types of deceitful cognitive manipulation. The first type operates through magical, or pseudomagical space-builders; others function by nonverbal means of deception—disguise, displacement, misplacement and replacement; still others employ verbal space-builders—lying, casuistry, and false promises.[28] The first chapter examines the exchange and substitution structures that underlie the "commerce" of sex; it highlights two problematic areas, the portrayal of women and the comic figure of the cuckold.

I do not propose a static classification of the different types of deception, such as that usually practiced by folklorists in search of "tale types"; I focus rather on how strategies of deception and the recurrent epistemic patterns that they generate contour the inner cognitive space of the *Contes* of La Fontaine. Such an analysis of the typology of deception in the *Contes* will lead to a more complete understanding of their inherent poetic composition.

I am conviced that the deceit patterns and the cognitive spaces they engender are an integral part of the poeticity of the *Contes*.[29] While there is no satisfactory theory of narrative poetry, poetry's innate self-referentiality and highly paradigmatic nature would make *all* repeated features, whether they be phonetic, semantic, or morphological, part of the poetic universe.

There is, in fact, something paradoxical about narrative poetry: narrative is linear and referential, whereas poetry is more synthetic and self-referential. Alain Vaillant sees a poem as producing in the reader an experience of simultaneously interacting components, much as a painting creates a unified perception for the viewer:

> La *mimésis* poétique a l'ambition, comme la *mimésis* picturale, de produire une représentation de la réalité; mais le poète, à la différence du peintre, ne peut disposer sur l'espace de la toile traits et couleurs, perceptibles d'un seul coup d'oeil: il lui faut le temps de la lecture, à condition que les mots ne s'évanouissent pas de la mémoire aussitôt qu'ils ont été lus. La *mimésis* poétique se reconnaît à cette tension particulière qui s'établit entre la linéarité du langage et la permanence artistique, vers laquelle elle tend sans jamais la réaliser pleinement; concrètement, c'est au nom de cette tension que le commentateur s'accorde le droit d'analyser un poème comme si tous les mots étaient présents, au même instant, dans la conscience du lecteur. (14)

In defining the poetic function of language, which is not exclusively confined to the genre of poetry, Roman Jakobson examines how linguistic signs are made palpable by the poetic functioning of language: "Comment la poéticité se manifeste-t-elle? En ceci que le mot est ressenti comme mot et non comme simple substitut de l'objet nommé, ni comme explosion d'émotion. En ceci, que les mots et leur syntaxe, leur forme externe et interne ne sont pas des indices indifférents de la réalité, mais possèdent leur propre poids et leur propre valeur" (124).[30] The Russian Formalists speak of poetry's "autotelicity": poetry (and art in general) is its own end.[31]

If the very shape and sound of words vibrate with the paradigms of interacting metaphors, rhythms, and semantic allusions, then repeated narrative patterns, as well, become palpable as an integral part of poetry. Narrative patterns occurring in the cognitive space of a poem take on their own paradigmatic importance. They are not only narratological "functions" (see Propp, *Morphologie du conte*), nor simple linear signifiers, pointing the way to the next step in the story. Exploiting the poetic function (in Jakobson's sense of the term), the deceit patterns inscribed in the text of the *Contes* are an integral part of La Fontaine's narrative poetics and possess a poetic value of their own in the same way metaphor, rhyme, and rhythms do. Figures, such as simile or anaphora, are not specific to poetry, and yet, when they appear in a poem they become integral to the poetic universe. In this way the patterned recurrences of deceit structures reverberate and interact with other poetic features of the *Contes*.

I hope that in my analysis of deceit structures I have not lost sight of a desire to arrive at a fuller appreciation of the *Contes* as poetry. "At any rate," states John C. Lapp, "there appears no doubt that far from being either 'base' or 'repugnantly facile,' the poetry of the *Contes* demonstrates an artistry equal to that of 'Adonis' or of the *Fables* themselves" (*Esthetics* 170).

NOTES

1. See my article, "Le Jeu de l'imitation: Un Aspect de la réception des *Contes* de La Fontaine."

2. For ample details on these editions and on the controversies they instigated, see Collinet's edition (1324-42), referred to from now on in the text as Collinet.

3. In the fabliaux, themes of seduction and adultery are prominent, but about two-thirds of the tales do not deal with such subjects—they "focus primarily on tricks and anti-social acts of varying kinds; many do not even have a male/female conflict" (Schenck 34).

4. The works listed are representative of recent research on the *Contes*. For a more complete listing see the bibliography in Sweetser's book and the bibliographical material regularly appearing in the journal published by La Société des Amis de Jean de La Fontaine, *Le Fablier*, especially in the 1991 issue.

5. See Wayne Booth, *The Rhetoric of Fiction*; Seymour Chatman, *Coming to Terms*, chapter 5; Gérard Genette, *Figures III*; Gerald Prince, *Dictionary of Narratology*, 42-43; Shlomith Rimmon-Kenan, *Narrative Fiction*, 86-89, 101-4.

6. In their *Dictionnaire encyclopédique* (1972) Ducrot and Todorov stated: "La catégorie du personnage est, paradoxalement, restée l'une des plus obscures de la poétique" (286). In 1976 Jonathan Culler added that "character is the major aspect of the novel to which structuralism has paid least attention and has been least successful in treating" (*Structuralist Poetics* 230). Two years later Seymour Chatman in *Story and Discourse* still found "remarkable how little has been said about the theory of character in literary history and criticism" (107). The paucity of critical interest in the fictional character has changed little since then, although notable exceptions are Chatman, *Coming to Terms* and Rimmon-Kenan, *Narrative Fiction*. Chatman calls for "a terminological distinction between these two loci of 'point of view': that of the narrator, and that of the character" (143).

7. On the cognitive dimension of the literary text, see Greimas, *Maupassant: La Sémiotique du texte* and the articles under "Cognitif" in Greimas and Courtés, *Sémiotique: Dictionnaire raisonné de la théorie du langage*.

8. On literature, film, and painting, see Fontanille, *Les Espaces subjectifs*; on film, Jost, *L'Oeil-caméra*; on linguistics, Nuyts, "Epistemic Modal Verbs and Adjectives and the Layered Representation of Conceptual and Linguistic Structure" and Kerbrat-Orecchionni, *L'Enonciation: De la subjectivité dans le langage*.

9. Fontanille explains why narratologists of the 1960s and 1970s were unable to construct a theory incorporating a cognitive agent: "les principales thèses narratologiques ont été conçues dans les années 60-70, après la grande fièvre cinémato-perspectiviste de l'après-guerre, et dans le contexte du structuralisme antimentaliste, où les accusations de 'psychologisme' fleurissaient comme des anathèmes; pas question à cette époque, d'admettre parmi les instances du récit un sujet cognitif" (*Les Espaces subjectifs* 37). He points out that Genette did admit, although hesitatingly, to the presence of a "sujet intermédiaire" (*Figures III* 179), but never really incorporated it into his theoretical framework.

10. See Hochman for an account of the assault on characters in the 1960s and 1970s (chap. 1, "The Case against Character," 13-27). Among the earlier group of semioticians, Philippe Hamon was the least reductive in his approach. See his schematization and analysis of *acteurs-actants* in "Pour un statut sémiologique du personnage."

11. For arguments on the complicated issue of characters as representations of people who, while not ontologically real, inhabit through fiction a possible world, see: W. H. Gass, *Fiction and the Figures of Life*; T. G. Pavel, *Fictional Worlds*; L. Pollard-Gott, "Attribution Theory and the Novel"; K. Walton, "How Remote are Fictional Worlds from the Real World?"; and R. Wilson, "The Bright Chimera: Character as a Literary Term."

12. Paul Ricoeur, in his analysis of the identification of the reader with the fictional character, finds the character's identity in a series of subjective transformations: "Si, en effet, toute histoire peut être considérée comme une chaîne de transformations qui conduisent d'une situation initiale à une situation terminale, l'identité narrative du personnage ne saurait être que le style unitaire de transformations subjectives réglées sur les transformations objectives qui obéissent a la règle de la complétude, de totalité et d'unité de l'intrigue" (292).

13. See these and other examples in R. W. Mitchell and N. S. Thomson, *Deception: Perspectives on Human and Nonhuman Deceit*; in particular, the articles by Lloyd on firefly communication, by Miles on deception among apes, and De Waal on deceptive communication among chimpanzees.

14. Chatman proposes the terms *slant* and *filter* to distinguish the "point of view" of the narrator from that of the character. His plea for new terminology is of capital importance:

> It is high time that we introduce a terminological distinction between these two loci of "point of view": that of the narrator and that of the character. I propose *slant* to name the narrator's attitudes and other mental nuances appropriate to the report function of discourse, and *filter* to name the much wider range of mental activity experienced by characters in the story world—perceptions, cognitions, attitudes, emotions, memories, fantasies, and the like. (*Coming to Terms* 143)

For Chatman *filter* still describes the mediating function of a character's consciousness. I believe that the concept of a cognitive *framing* of experience within the diegetic world is more appropriate to my analysis of deceit, which focuses on the characters as agents with discrepant states of knowledge.

15. Narratologists, such as Genette, Mieke Bal, or Rimmon-Kenan, who use such terms are chiefly interested in analyzing a center of consciousness mediating the story.

16. See my analysis of cognitive irony in "Ironie dramatique ou ironie cognitive?"

17. Gérard Genette's terminology (*Figures III*) concerning narrative levels (diegetic, extradiegetic, and metadiegetic) is precise and particularly apt for my purposes. These terms will recur throughout my study.

18. Wayne C. Booth's discussion of the scene in *Tartuffe* (66-67) does not analyze the audience's perception of irony: "The audience need not be subtle; they need only be awake and morally engaged and every ironic touch will be evident."

19. See the analysis and diagrams under "manipulation" in both volumes of Greimas and Courtés, *Dictionnaire*, and in Fontanille, *Le Savoir partagé*, 38-42.

20. For an overview of the chief theorists on the subject, see *Framing in Discourse*, edited by Deborah Tannen, particularly the introduction to her own chapter, "What's in a Frame? Surface Evidence for Underlying Expectations," where she categorizes the main theorists according to their disciplines, and their choice of the following terms: schema, script, and frame.

21. On the concept of mental space in terms of cognitive science and information theory, see Dinsmore and Fauconnier.

22. See the canonical text on the subject, Goffman's *Frame Analysis*. Gregory Bateson initiated the term "frame" and describes frames as metacommunication, "communication about communication" (209). Ludwig Wittgenstein sees them as the description under which an event is to be seen (198, 202). Manning and Hawkins in *Beyond Goffman* see a frame as providing the rules and principles that lead to an understanding of experienced events: "A frame addresses the question 'What is going on here?'" (207).

23. For a taxonomic approach to the arrangement of these frames into patterns of deceit, see the Conclusion and Appendix B.

24. All references to the *Contes* are to Collinet's 1991 edition in the Pléiade series.

25. On closure in La Fontaine's *Fables*, see the perceptive analysis of Richard Danner in his chapter, "Irony and Closure" (64-95).

26. In the bibliography of *conteurs* published at the end of my article, "Le Jeu de l'imitation: Un Aspect de la réception des *Contes* de La Fontaine," the entries reveal that the most common generic expression used by the authors themselves in the titles of their works is *contes en vers*. Indicative of the hesitation over the nomenclature, however, the 1905-06 Van Bever anthologies of licentious tales in verse were entitled: *Les Conteurs libertins du XVIIIe siècle*, *Conteurs galants du XVIIIe siècle*, and *Contes et conteurs gaillards au XVIIIe siècle*. Folklorists commonly use the term *contes en vers* to describe this body of eighteenth- and nineteenth-century libertine tales—see G. Legman, "Toward a Motif-Index of Erotic Humor" (236-38) and my article, "The *Conte en vers*: Expanding Stith Thompson's X-File of Obscene Motifs."

27. If certain poems in the *Contes et nouvelles en vers* can be excluded from the genre as such, then, one might ask, could some poems published in the *Fables* not be considered *contes en vers*? After all, there are lengthy fables, there are fables without explicit moral lessons, and there are even fables that tell tales about humans. I argue that the specificity of the *conte en vers*, as compared to the fable genre, involves more than the length of the tale, the lack of an explicit moral lesson, and the presence of humans as opposed to animals. The specificity

of the *conte en vers* genre precludes extending the genre, either to include animal fables or those fables having human characters, but which have neither an erotic dimension (understood minimally to include a male-female relationship) nor an element of deception in the narrative. Self-deception is present, for instance, in "La Laitière et le pot au lait," but the licentious component is absent. Such fables quite plainly do not belong to the genre of the *conte en vers* because the sexual dimension is lacking, although they are a type of *conte*. Aside from the problematic "La Matrone d'Ephèse" which has found its way at different times into both the *Fables* and the *Contes*, six fables do, in varying degrees, resemble the *conte en vers* genre: "L'Ivrogne et sa femme" (III, 7), "La Femme noyée" (III, 16), "La Jeune Veuve" (VI, 21), "Le Mal Marié" (VII, 2), "Les Femmes et le secret" (VIII, 6), and "Le Mari, la femme, et le voleur" (IX, 15). The first two, however, center on a *facétie verbale*; in the next two dupery is not a prominent element; the last two approximate most closely the *conte en vers* as it was practiced by La Fontaine in the *Contes et nouvelles en vers*.

28. Mary Jane Stearns Schenck, following a Proppian methodology (Propp, *Morphologie du conte*) analyzes 66 self-proclaimed fabliaux by attempting to isolate functions from the actions. Her fifth function is deception under which she enumerates six different forms: lying or fabricating a story, hiding, disguising/substituting, cheating, playing on words/riddling, using magic.

29. On the *Contes* as poetry, see John C. Lapp, "The Esthetics of Negligence: La Fontaine's *Contes*" and *The Esthetics of Negligence* (159-70), as well as Collinet, "La Fontaine est-il poète?"

30. I am aware that some do not accept into the category of true poetry a poem possessing the degree of narrativity present in the *Contes*. And yet, in French poetry before Baudelaire, poetry and narrativity were not mutually exclusive. The attempt to reduce poetry to the lyric genre has not been wholly successful: "Il semble que, très récemment, depuis les années 1970-1980, l'exclusion qui frappait le narratif, le descriptif et le didactique de la poésie tende à s'atténuer. Le 'terrorisme' contre le récit, en particulier, ne semble plus aujourd'hui de mise en France, de sorte que le rêve d'une 'poésie pure' paraît avoir fait long feu" (Combe 73).

31. On Sklovskij's discussion of the palpableness of poetic language and of the autotelicity of poetry, see Ejxenbaum, "The Theory of the Formal Method," 12-14.

CHAPTER ONE

EXCHANGE AND SUBSTITUTION: SEXUAL COMMERCE IN THE *CONTES*

The *Contes* of La Fontaine belong to the same literary tradition as farce, fabliau, and the Renaissance novella, genres representing in a comic mode actions that disrupt accepted social conventions. These *genres facétieux* with their characteristic portrayal of deviousness and deception are frequently the purveyors of antibourgeois satire. They are narrative genres depicting subversive actions that, whether they be a misdeed such as cheating in a business transaction or infidelity in marriage, upset the social equilibrium, undermining established exchange patterns, or what could be termed the "commerce" among individuals in society.[1] A passage in La Fontaine's tale, "La Coupe enchantée," is indicative of the close affiliation of the *Contes* with the *facétieux* genre and its comic treatment of unconventional practices:

> Laissez les bons bourgeois se plaire en leur ménage;
> C'est pour eux seuls qu'Hymen fit les plaisirs permis.
> Mais vous! ne pas chercher ce qu'Amour a d'exquis!
> Les plaisirs défendus n'auront rien qui vous pique!
> Et vous les bannirez de votre république!
> Non, non, je veux qu'ils soient désormais vos amis.
> (ll. 234-39)

The principle of exchange is basic to human society. The barter system, the justice system, marriage, social contracts, the monetary system, and political negotiations all rely on the principle of exchange. French literary theory in the early 1970s was highly indebted to the theories of Marx, Levi-Strauss, Saussure, and Freud on reciprocity, gift exchange, and the circulation of money, women, and goods. This approach is best exemplified in the works of Barthes and Greimas.[2] Over the past twenty years the problematic of exchange has made its way into the mainstream of literary theory.[3]

The *Contes* of La Fontaine nearly all represent some form of sexual "commerce" and constitute an ideal corpus for the analysis of narratives of deceit with their attendant patterns of exchange and substitution.

In an equitable exchange the goods traded are considered to be equal in value. When the principle of equivalency is contravened by one of the parties through a form of devious substitution, deceit occurs and a subversion of the exchange is accomplished. In some instances, only the deceivee's part of the exchange takes place, with no reciprocity at all on the part of the deceiver. This is what happens in La Fontaine's tale, "A Femme avare galant escroc," where Gulphar tricks a woman into adultery and cleverly manages not to pay the agreed sum of money to her. To use an example from another context, the same pattern of deceit occurs when ransom money is duly handed over, but the kidnappers do not return the hostage. A second duplicitous stratagem consists in secretly making a fake substitution that, were it known beforehand, would be unacceptable to the other party of the exchange—the ransom money handed over to the kidnappers turns out to be counterfeit. In "Richard Minutolo" a woman makes a deal with Richard to substitute herself for her husband's mistress at a supposed rendezvous, only to discover—too late—that it is Richard himself with whom she is making love. A third deceitful exchange is the promise of spurious goods to the victim of the trickery. A monk promises holiness to a young girl if she consents to sex with him in "Le Diable en enfer." The latter instance of dupery is not unlike the situation occurring when small investors are taken in by the promise of high interest rates if they invest in what turns out to be a nonexistent company. And finally, perhaps the ultimate deception is to make a substitution when the deceivee is unaware even that an exchange is taking place. In "La Servante justifiée" the wife replays the love scene in the garden with her husband, not knowing that he is only covering up his former escapade in the same place with the female servant.

Many variations are played on these basic structures. Whether the trickery in the *Contes* takes on the guise of magic, of disguise, the usurpation of space (displacement, misplacement, replacement), or of lying, false promises, or casuistry, as we shall see in the following chapters these all, in one way or another, constitute patterns of subverting an exchange process.

THE COMMERCE OF SEX AND MONEY

The association of sex and money, so prevalent in the fabliau tradition, was also reflected in medieval penitential handbooks and treatises on usury in which producing money from money was seen as spiritual fornication: *spiritualis fornicatio incorrigibilis*.[4] Dante places usurers in the circle of Soddoma, among those with unnatural sexual practices. While La Fontaine found in his sources ample material for regarding sexual relationships as a kind of commercial transaction, he frequently invents or amplifies monetary references and analogies, often moving into the satiric mode and commenting on the *moeurs* of seventeenth-century France.[5] The age of *préciosité* is dead; the voyage of love no longer passes through the village of "Soupirs":

> Ce n'est pas chose en ce siècle fort rare.
> Je l'ai jà dit, rien n'y font les soupirs.
> Celui-là parle une langue barbare,
> Qui l'or en main n'explique ses désirs.
> ("A Femme avare galant escroc," ll. 26-29)

Even the language of love has been supplanted by the eloquence of money:[6]

> Le compagnon ne put rien tirer d'elle,
> Qu'il ne parlât. Chacun sait ce que c'est
> Que de parler: le lecteur s'il lui plaît
> Me permettra de dire ainsi la chose.
> Gulphar donc parle, et si bien qu'il propose
> Deux cents écus. La belle l'écouta.
> (ll. 36-41)

This tale concludes with a line that makes the reference to French society explicit: "Ainsi vit-on chez nous autres François" (l. 71). Money as the rhetoric of love is the theme of a long development, wholly of La Fontaine's invention, at the end of "Pâté d'Anguille" where money ("mots dorés") is presented as being more persuasive than the orators of antiquity:

> On dit qu'il parlait comme un ange,
> De mots dorés usant toujours:
> Mots dorés font tout en amours.
> C'est une maxime constante:
> Chacun sait quelle est mon entente:

J'ai rebattu cent et cent fois
Ceci dans cent et cent endroits:
Mais la chose est si nécessaire
Que je ne puis jamais m'en taire,
Et redirai jusques au bout
Mots dorés en amours font tout.
Ils persuadent la donzelle,
Son petit chien, sa demoiselle,
Son époux quelquefois aussi;
C'est le seul qu'il fallait ici
Persuader; il n'avait l'âme
Sourde à cette éloquence; et dame
Les orateurs du temps jadis
N'en ont de telle en leurs écrits.
(ll. 115-33)

When Fédéric, in "Le Faucon," tries to win over Clitie, offering her money is a strategy that is esteemed more powerful than the persuasive words of Cicero:

S'agissait-il de divertir la dame,
A pleines mains il vous jetait l'argent:
Sachant très bien qu'en amour comme en guerre,
On ne doit plaindre un métail qui fait tout;
Renverse murs; jette portes par terre;
N'entreprend rien dont il ne vienne à bout;
Fait taire des chiens; et quand il veut servantes;
Et quand il veut les rend plus éloquentes
Que Cicéron, et mieux persuadantes.
(ll. 10-18)

An excursus deploring the power of money in the sexual transaction breaks through the narrative in "La Coupe enchantée":

L'argent en aurait fait l'affaire.
Et quelle affaire ne fait point
Ce bienheureux métail l'argent maître du monde?
Soyez beau, bien disant, ayez perruque blonde,
N'omettez un seul petit point;
Un financier viendra qui sur votre moustache
Enlèvera la belle; et dès le premier jour

Il fera présent du panache;
Vous languirez encor après un an d'amour.
(ll. 311-19)

In "Les Quiproquos" relationships between men and women, once a social *commerce* have now become a monetary *commerce*:[7]

Las! ce n'est plus le siècle de nos pères.
Amour vend tout, et nymphes et bergères;
Il met le taux à maint objet divin:
C'était un dieu, ce n'est qu'un échevin.
O temps! ô moeurs! ô coutume perverse!
(ll. 77-81)

The commerce of sex places on what the woman has to offer in the exchange a value equivalent to money, and her offering is alluded to metaphorically as money. In "Le Petit Chien qui secoue de l'argent et des pierreries," after Favori, the magic dog, has produced an abundance of money and precious jewels, the pilgrim explains the exchange of money for sex with Argie in the following terms:

C'est un prodige enfin: madame cependant
En a comme on dit la monnoie.
Pourvu que j'aie cette joie
De coucher avec elle une nuit seulement
Favori sera sien dès le même moment.
(ll. 218-22)

Argie's *confidente* understands the proposed transaction in similar fashion, alluding to Argie's "trésor":

Comment peut-on refuser rien?
Puis celui-ci possède un chien
Que le royaume de la Chine
Ne paierait pas de tout son or:
Une nuit de madame aussi c'est un trésor.
(ll. 235-39)

Later on when Argie explains her motivation to her husband, she depicts herself as goods belonging to him and the exchange as an astute transaction taking place between a "donneur" and a woman who knows how to manage her husband's merchandise:

> C'est de ce chien qu'on m'a fait don.
> Il a bâti cette maison.
> Puis faites-moi trouver au monde une Excellence,
> Une Altesse, une Majesté,
> Qui refuse sa jouissance
> A dons de cette qualité;
> Surtout quand le donneur est bien fait, et qu'il aime,
> Et qu'il mérite d'être aimé.
> En échange du chien l'on me voulait moi-même;
> Ce que vous possédez de trop, je l'ai donné;
> Bien entendu monsieur; suis-je chose si chère?
> Vraiment vous me croiriez bien pauvre ménagère
> Si je laissais aller tel chien à ce prix-là.
> (ll. 473-85)

There are two predominant metaphors for love in La Fontaine's *Contes*. One is love as a *jeu*; the other is love as a *marché*.[8]

WOMAN AS A CONSUMABLE COMMODITY

In the exchange or bartering context a woman is the principal object of value to be sought after, to be purchased, to be traded, and to be consumed. Jane Merino-Morais's semiotic analysis of "La Fiancée du roi de Garbe" demonstrates how the very structures of the narrative are determined by the ideology of the woman as object of male desire: "[L]a fiancée de notre conte se présente en tant qu'objet convoité qui passe de main en main" (11).

Nevertheless, I argue that the *Contes* is not a strongly misogynist text. Just as a reassessment of the fabliau as a misogynist genre is now underway,[9] it is time for a closer look at how women are viewed in the *Contes*. Could the following lines from "Les Rémois" deploring the status of women as no more than a prey and plaything for men come from the pen of an unconditional misogynist?

> Femmes voilà souvent comme on vous traite.
> Le seul plaisir est ce que l'on souhaite.
> Amour est mort: le pauvre compagnon
> Fut enterré sur les bords du Lignon.

Nous n'en avons ici ni vent ni voie.
Vous y servez de jouet et de proie
A jeunes gens indiscrets, scélérats:
C'est bien raison qu'au double on le leur rende:
Le beau premier qui sera dans vos lacs,
Plumez-le-moi, je vous le recommande.
(ll. 48-57)

It is true that a misogynist perspective haunts La Fontaine's text and surfaces with particular vehemence in some tales. The *Contes* retell many stories about women who are victimized. Men like Joconde go off seeking women to "conquer." Richard Minutolo takes a perverse pleasure in making love to a woman who thinks he is her husband. Stupid or naive women are exploited in tales like "Comment l'esprit vient aux filles." In one tale, "Clochette," a thirteen-year-old girl is raped by force.

On the other hand, the text presents women treating men, especially older men, as objects of derision to be manipulated and discarded with impunity. Aminte's husband Géronte is typical of the older husband, "malgracieux, incommode et jaloux" ("La Confidente sans le savoir ou le Stratagème," l. 22); he is fit to be replaced. Such men are depicted as *sots* who deserve their wives' infidelity. Young innocent men who become the sexual prey of women are not absent either from the world of the *Contes*. Chosen as a lover more for his beauty than his intelligence, Nicaise is abruptly dismissed as a "prince des sots" (l. 186) by a woman who despises him for being more interested in keeping her clothes unsullied than in making love to her. The nuns in "Mazet de Lamporechio" are quite ready to indulge in the sexual exploitation of the handsome but supposedly stupid and mute gardener.

Women, far from being victims, frequently assume the function of deceivers in the *Contes*. In well over half of the tales depicting a male-female relationship, a woman is the agent of the trickery. For Marc Soriano, even "La Chose impossible," a tale evoking the medieval linkage of woman with the devil, is not antifeminist because La Fontaine treats the motif of the 'fil' (a pubic hair) as a positive sign: "L'obstination, figurée et symbolisée par 'l'annelure' qui résiste à tous les traitements, joue cette fois un rôle salvifique, puisque grâce à elle, la femme se montre plus rusée que le diable et sauve l'âme de son amant" (104-5). Women are represented as agents of their own desire and are praised for their cleverness. "La Confidente sans le savoir ou le stratagème" concludes with a eulogy of Aminte's intelligence:

> Mais, dites-moi, de grace, je vous prie,
> Qui vous a fait aviser de ce tour?
> Car jamais tel ne se fit en amour.
> Sur les plus fins je prétends qu'il excelle;
> Et vous devez vous-même l'avouer.
> Elle rougit et n'en fut que plus belle;
> Sur son esprit, sur ses traits, sur son zèle,
> Il la loua; ne fit-il que louer?
> (ll. 191-98)

What is more, most of the women who are objects of male deceit are represented as becoming willing partners in the sexual capers into which the men lead them. This spirited amoral approach is typical of the *Contes*, where both men and women cavort in the sexual arena.

Critics of La Fontaine's *Fables* are adept at exploring the implications of the narrator's discourse for the interpretation of individual texts;[10] divergent features are analyzed because, as David Lee Rubin, points out: "the sense of a given fable by La Fontaine is never to be taken for granted" (50). The same approach must be applied to the *Contes* where already textualized narratives are retold in a wholly new discursive context so that frequently—as in the *Fables*—the ideological link with the original becomes tenuous. The narrator's voice in the *Contes* sometimes functions as a foil to the misogyny of the source-tales, attenuating circumstances, praising female resourcefulness, or adding commentary on the woman's viewpoint and reactions. The author-narrator demonstrates awareness of criticism directed at him for his portrayal of women. In the 1665 Preface La Fontaine adopts the ever-common, but unsatisfactory, defense that any antifeminism is not intended *seriously*:

> Quant à la seconde objection, par laquelle on me reproche que ce livre fait tort aux femmes, on aurait raison si je parlais sérieusement; mais qui ne voit que ceci est jeu, et par conséquent ne peut porter coup? Il ne faut pas avoir peur que les mariages en soient à l'avenir moins fréquents, et les maris plus forts sur leurs gardes. (Collinet 557)

In the preamble to "Les Oies de frère Philippe" La Fontaine envisages his covert female readers and justifies their reading the *Contes* as a harmless activity:

> Je dois trop au beau sexe; il me fait trop d'honneur
> De lire ces récits; si tant est qu'il les lise.

> Pourquoi non? c'est assez qu'il condamne en son coeur
> Celles qui font quelque sottise.
> Ne peut-il pas, sans qu'il le dise,
> Rire sous cape de ces tours,
> Quelque aventure qu'il y trouve?
> . . . .
> Beau sexe, vous pouvez le lire en sûreté.
> (ll. 1-7, 40)

In "Le Fleuve Scamandre" La Fontaine resorts to the classic defense of morally dubious material in a literary work by arguing that exemplary lessons derive from the depiction of vice:

> Si, dans ces vers, j'introduis et je chante
> Certain trompeur et certaine innocente,
> C'est dans la vue et dans l'intention
> Qu'on se méfie en telle occasion:
> J'ouvre l'esprit, et rend le sexe habile
> A se garder de ces pièges divers.
> Sotte ignorance en fait trébucher mille,
> Contre une seule à qui nuiraient mes vers.
> (ll. 10-17)

Although the author-narrator is not unaware of the female reader, the implicit reader remains male, and a particularly obtuse male when it comes to understanding women. The reader's assumed ignorance motivates the narrator's claim to special competence in explaining the psychology of women. For example, in "Joconde" just following the tender farewell scene between the husband and wife, the narrator cautions the reader:

> Vous autres bonnes gens eussiez cru que la dame
> Une heure après eût rendu l'âme;
> Moi qui sais ce que c'est que l'esprit d'une femme,
> Je m'en serais à bon droit défié.
> (ll. 84-87)

On another occasion, in "La Mandragore," the narrator explains to the male reader, represented in the text through a dialogue, why Lucrèce dresses so well to go to bed with a miller. The uncomprehending reader, having just read that Lucrèce was "ce soir atournée" (l. 224), exclaims: "Voire, ce soir? atournée; et pour qui?" (l. 225). The narrator patiently addresses the male reader's problem, repeating the question, demonstrating that he understands it, and then answering it, all the while assuming the role of one who comprehends women:

> Pour qui? j'entends: n'est-ce pas que la dame
> Pour un meunier prenait trop de souci?
> Vous vous trompez; le sexe en use ainsi.
> Meuniers ou rois, il veut plaire à toute âme.
> C'est double honneur, ce semble en une femme,
> Quand son mérite échauffe un esprit lourd,
> Et fait aimer les coeurs nés sans amour.
> (ll. 226-32)

The *Contes* even attempt in some passages to represent female subjectivity and female desire, a dimension that was minimal in the texts of both Ariosto and Boccaccio.[11] In "Le Roi Candaule et le Maître en droit" the narrator's determination to understand and communicate to the reader a woman's point of view results in the following lines where he wishes the male reader could become a woman so that he would understand the queen's angry and vengeful state of mind upon discovering that her husband has let another man view her taking a bath:

> Je voudrais pour un moment,
> Lecteur, que tu fusses femme:
> Tu ne saurais autrement
> Concevoir jusqu'où la dame
> Porta son secret dépit.
> (ll. 74-78)

Then, through free indirect discourse, the queen's thoughts are evoked:

> Un mortel eut le crédit
> De voir de si belles choses,
> À tous mortels lettres closes!
> (ll. 79-81)

The narrator goes on to explain further the queen's interior rage:

> Ces pensers incitaient la reine à la vengeance.
> Honte, dépit, courroux, son coeur employa tout.

Amour même, dit-on, fut de l'intelligence:
De quoi ne vient-il point à bout?
(ll. 86-89)

The representation of female sexual pleasure is a further demonstration of La Fontaine's adaptation of the *conte en vers* genre for a seventeenth-century French readership where women were a vital component of the literary world. In "Le Berceau" the wife is depicted enjoying her experience with a man whom she believes to be her husband:

Qu'a mon mari, dit-elle, et quelle joie
Le fait agir en homme de vingt ans?
Prenons ceci, puisque Dieu nous l'envoie;
Nous n'aurons pas toujours tel passe-temps.
(ll. 117-20)

In "Le Calendrier des vieillards" La Fontaine takes a position on women's right to sexual pleasure when he satirizes a husband who treats his wife as a passionless doll, providing her with everything but sex. The adverb "petitement," highlighted by an enjambment, encapsulates Bartholomée's sad plight:

Notre docteur régalait sa moitié,
Petitement; enfin c'était pitié.
A cela près, il traitait bien sa femme.
Les affiquets, les habits à changer,
Joyaux, bijoux, ne manquaient à la dame;
Mais tout cela n'est que pour amuser
Un peu de temps des esprits de poupée.
(ll. 82-88)

After Bartholomée has spent two days with the pirate Pagamin she makes a blunt declaration to her miserly husband; another enjambment evokes the erotic space between *tête* and *talon*:

J'ai plus appris des choses de la vie
Depuis deux jours, qu'en quatre ans avec vous.
Laissez-moi donc, monsieur mon cher époux.
. . . .
Adieu vous dis, vous et vos jours de fête.
Je suis de chair, les habits rien n'y font:

Vous savez bien, monsieur, qu'entre la tête
Et le talon d'autres affaires sont.
(ll. 218-20, 238-41)

The erotic "female gaze" may be a male construct in the *Contes*, but it is, nevertheless, legitimized. The widow who seduces Renaud in "L'Oraison de saint Julien" looks on him with desire:

Quant à la veuve, elle ne fit en somme
Que regarder, témoignant son désir:
Soit que déjà l'attente du plaisir
L'eût disposée; ou soit par sympathie;
Ou que la mine, ou bien le procédé
De Renaud d'Ast eussent son coeur touché.
(ll. 222-27)

Anne in "Le Cas de conscience" takes enormous pleasure in contemplating the nude body of a young man who is bathing in a river:

Elle vit un jeune garçon
Se baigner nu. La fillette était drue,
Honnête toutefois. L'objet plut à sa vue.
Nuls défauts ne pouvaient être au gars reprochés:
. . . .
Ça et là ses regards en liberté couraient
Où les portait leur fantaisie,
Ça et là, c'est-à-dire aux différents attraits
Du garçon au corps jeune et frais,
Blanc, poli, bien formé, de taille haute et drète,
Digne enfin des regards d'Annette.
(ll. 37-40, 46-51)

Her gaze is interrupted for a few lines as scruples invade her consciousness (ll. 55-60); La Fontaine draws an analogy between the activity of artists sketching a nude and the visual impression made on the memory of Anne as she rivets her eyes with renewed intensity on the pleasing spectacle before her:

Annette la contemplative
Regarda de son mieux. Quelqu'un n'a-t-il point vu
Comme on dessine sur nature?

On vous campe une créature,
Une Eve, ou quelque Adam, j'entends un objet nu;
Puis force gens assis comme notre bergère
Font un crayon conforme à cet original.
Au fond de sa mémoire Anne en sut fort bien faire
Un qui ne ressemblait pas mal.
(ll. 64-72)

The text of La Fontaine's *Contes* brings into prominence in a satirical light a number of social problems affecting women in the seventeenth century. The strongest castigation of arranged marriages in which the woman is treated as a commodity to be exchanged between her parents and the future husband occurs in "Le Calendrier des Vieillards."[12] The tale opens with the following social commentary:

Plus d'une fois je me suis étonné
Que ce qui fait la paix du mariage
En est le point le moins considéré,
Lorsque l'on met une fille en ménage,
Les père et mère ont pour objet le bien;
Tout le surplus, ils le comptent pour rien;
Jeunes tendrons à vieillards apparient.
(ll. 1-7)

In the midst of the narrative the theme recurs when Bartholomée berates her husband and her parents for marrying her against her will:

. . . s'en est-on mis en peine
Quand malgré moi l'on m'a jointe avec vous?
Vous vieux penard, moi, fille jeune et drue,
Qui méritais d'être un peu mieux pourvue,
Et de goûter ce qu'hymen a de doux.
. . . .
Vous et les miens avez mérité pis.
Vous pour avoir mal mesuré vos forces
En m'épousant; eux, pour s'être mépris,
En préférant les légères amorces
De quelque bien à cet autre point-là.
(ll. 205-9, 224-28)

The narrator advises young women against premarital sex strictly from a practical standpoint, since most men are not understanding and certainly not many are as gullible as the "roi de Garbe":

> Filles maintenez-vous; l'affaire est d'importance;
> Rois de Garbe ne sont oiseaux communs en France;
> Vous voyez que l'hymen y suit l'accord de près.
> ("La Fiancée du roi de Garbe," ll. 787-89)

However, to the unmarried woman who strays from the path of virtue, La Fontaine offers a sympathetic and practical recommendation:

> Si quelqu'une pourtant ne s'en pouvait défendre,
> Le remède sera de rire en son malheur.
> Il est bon de garder sa fleur;
> Mais, pour l'avoir perdue, il ne se faut pas pendre.
> (ll. 798-801)

In another tale, the narrator comments in an aside that lying about one's virginity is legitimate: "Fille qui ment ne pèche pas" ("Nicaise," l. 228). The same tactic is urged in the case of children born previous to marriage or of infidelity in marriage: "Mentir alors est digne de pardon" ("Les Aveux indiscrets," l. 115). The narrator's practical, disabused view of marriage as a reciprocal venture voices itself in "Belphégor":

> Mais, dira-t-on, n'est-il en nulles guises
> D'heureux ménage? Après mûr examen,
> J'appelle un bon, voire un parfait hymen,
> Quand les conjoints se souffrent leurs sottises.
> (ll. 121-24)

The *Contes*, then, are not essentially misogynist tales that represent woman as an object of exchange among males. Admittedly, La Fontaine does not discard the antifeminism of his sources completely; sometimes what his narrator appears to view as seduction does not involve a woman's full consent.[13] And yet, woman is not presented uniquely as an object of male consumption. As he states in the 1665 Preface, the *Contes* aim at presenting in a ludic atmosphere ("Ceci est jeu") the "bons tours des maris et des femmes" ("Joconde"). These tales of men and women who subvert the social order with impunity unfold under the amused and admiring eye of an author-narrator who is more concerned about diverting the reader than about taking a stance in the *querelle des femmes*.

I would even venture to say that compared to the concerns of "la belle Arthénice" and the *précieuse* society she represents—a group generally considered more favorably than La Fontaine for its feminism—La Fontaine's discussions of the real problems of women are more genuinely "feminist" than the *questions d'amour* debated in the ranks of the *précieuses*.

CUCKOLDRY: MAN AS VICTIM OF SEXUAL COMMERCE

Cuckoldry, an ancient topos and a commonplace of European Renaissance literature, is one of the prominent themes of La Fontaine's *Contes*. Cuckoldry, at least as it came to be viewed in literature, is a masculine fantasy based on a patriarchal concept of marriage.[14] In a patriarchal society a woman's infidelity is a serious offense to her husband because his rights of exclusive possession—his territorial rights over his wife—are infringed upon when she is unfaithful. Her transgression reveals his incapacity to exert control over his "property."[15] Her act of infidelity also diminishes her intrinsic value, as Keith Thomas points out in his article on "The Double Standard": "[T]he value of this property is immeasurably diminished if the woman at any time has sexual relations with anyone other than her husband" (210).

Even in those cases where someone, like La Fontaine's "Muletier," makes love to a man's wife by tricking her into thinking he is her husband, the damage to the marriage contract is simply that she, willingly or not, has had a sexual encounter with someone else. That makes her unchaste and less valuable to her husband; his honor therefore suffers. The focus for blame is not "the other man"; the woman is blamed: "Nos femmes, ce dit-il, nous en ont donné d'une" ("Joconde," l. 244). Women are the betrayers in the adulterous exchange and men are the victims. A wife can be betrayed by her husband, but she, unlike him, is not looked upon as an object of pity or humor. Her infidelity casts shame on her husband in a world that by virtue of the double standard tolerates a man's infidelity, while condemning a woman's.[16]

Given this social situation, why does the victimized husband become a comic figure in the literary world of cuckoldry? And why is the wife most often cheered on in her dupery rather than being condemned? The comic effect can be explained in part by the depiction of the wronged husband as a fool whose inflated ego deserves ridicule. The husband's pride suffers not only from the rebuff to his sense of ownership, but also because a wife's substitution of another man as a sexual partner involves on her part an unfavorable judgment of her husband. Cuckoldry is a wife's treachery and consists in substituting someone else in the place belonging to her husband.[17] The unstated message in the wife's infidelity is that the substitute is valued more highly than the husband because he possesses more desirable qualities, such as youth, fortune, or beauty. Through her action the cuckold undergoes a sudden and uncontrollable loss of dignity; his vastly overblown sense of self-importance is unceremoniously deflated. The often pompous and ineffectual husband suffers comic humiliation because he over-esteems his own worth and his own invulnerability. Inevitably, he slips on the banana peel of his own vanity and provokes the spectator's laughter.

Another source of the comic register is the enormous disparity between the cuckold's perception of his capabilities and his obliviousness to the problematic nature of his situation. The reader enjoys the comic double vision of how the cuckold views himself and of how others view him. Like the monomaniacal characters of Molière's comedies, the betrayed husband is represented by La Fontaine as a supreme *sot*:[18]

> Figurez-vous un fou chez qui tous les soupçons
> Sont bien venus, quoi qu'on lui die.
> Il n'a pas un moment de repos en sa vie.
> Si l'oreille lui tinte, ô dieux! tout est perdu.
> Ses songes sont toujours que l'on le fait cocu.
> . . . .
> Il veut à toute force être au nombre des sots,
> Il se maintient cocu, du moins de la pensée,
> S'il ne l'est en chair et en os.
> ("La Coupe enchantée," ll. 3-7, 17-19)

Incapable of seeing and especially of knowing what is going on in his entourage, the *sot* typically takes exceptional measures in his efforts to see and to know about the actions of his wife:

> Le malheureux époux s'informe, s'inquiète,
> Et de tout son pouvoir court au-devant d'un mal
> Que la peur bien souvent rend aux hommes fatal.
> ("La Coupe enchantée," ll. 369-71)

The jealous husband focuses on keeping his wife chaste; he goes to great lengths to ensure this:

> Il captivait sa femme cependant;
> De ses cheveux voulait savoir le nombre;

La faisait suivre, à toute heure, en tous lieux,
Par une vieille au corps tout rempli d'yeux,
Qui la quittait aussi peu que son ombre.
("On ne s'avise jamais de tout," ll. 8-12)

The reader decodes the disparity between the character's view of reality and his actual situation of impotency as a sign of the comic register.

In comic genres the serious side of infidelity is suspended in favor of the delightful spectacle of a wife who outmaneuvers her husband. A woman's lack of power in this masculine-dominated society contributes to make of her role a significant factor in inciting a comic response. Ironic laughter comes easily when the underdog temporarily upsets the status quo; for a moment there is a pleasurable and giddy glimpse of a *monde renversé*. In the world of the *Contes* male infidelity is acceptable. Men are the accepted *trompeurs*; when a woman is empowered to enter this forbidden area of sexuality, she takes on the comic role of deceiving the deceiver, who becomes the *trompeur trompé*. Her success in reversing the social setup and victimizing her husband brings to the reader a view of the alternate world of comedy.

The "epistemological disquiet"[19] inherent in the theme of cuckoldry harbors great potential for exploitation on the cognitive level of the text. Through his narrator La Fontaine muses playfully on the status of the *cocu* as the text explores a variety of situations in the devious and manipulative world of cuckoldry. Can a man be a cuckold without knowing it? Can a woman be unfaithful if she believes erroneously that the man she is making love with is her husband? Conversely, does a man become a cuckold if his wife makes love to him believing he is someone else? Such are some of the dilemmas posed in La Fontaine's *Contes*.

In "La Mandragore" Callimaque, the would-be lover, explains—with a dash of casuistry— to Nicia, the foolish husband that they can arrange for his wife to sleep with another man without Nicia becoming a *cocu*, so long as his wife does not enjoy it and her partner does not know who she is:

Votre moitié n'ayant lieu de s'y plaire,
Et le coquin même n'y songeant pas,
Vous ne tombez proprement dans le cas
De cocuage. . . .
(ll. 160-63)

The same argument of insufficient knowledge is made at the end of "Les Quiproquos" where the husband, Clidamant, sends his friend to an assignation, not knowing that his own wife has usurped the female servant's place, believing her husband will be there:

> . . . J'ai vu les gens mouvoir
> Deux questions; l'une, c'est à savoir
> Si l'époux fut du nombre des confrères,
> À mon avis n'a point de fondement,
> Puisque la dame et l'ami nullement
> Ne prétendaient vaquer à ces mystéres.
> (ll. 185-90)

In "La Coupe enchantée" Damon, through a magical enchantment, takes on the appearance of someone else and manages to persuade his wife, Caliste, to kiss him. Knowing that she would shortly succumb completely, he stops the charade to muse on the problem of whether he is now truly a *cocu*:

> Un cocu se pouvait-il faire
> Par la volonté seule et sans venir au point?
> L'était-il, ne l'était-il point?
> (ll. 341-43)

The patriarchal exchange pattern would dictate a male-to-male transfer of a woman, whether from father to husband in the marriage contract or from husband to lover by way of usurpation in adultery. In the comic world of cuckoldry, as it is presented in La Fontaine's *Contes*, that process is short-circuited by the wife who subverts the pattern and in so doing turns the comic gaze on the victimized husband.

"JOCONDE": A WORLD OF CUCKOLDRY AND COGNITIVE ILLUSIONS

The liminary tale of the *Contes*, "Joconde," illustrates how sexual commerce with all its possibilities for cognitive convolutions is exploited textually by La Fontaine. The source of "Joconde" is Ariosto's *Orlando furioso* (canto XXVIII, 1-74) where narrating the tale of Iocondo serves an exemplary purpose by assuaging the grief of Rodomonte who has just lost his beloved Doralice to Mandricardo, his rival. Ariosto's work was well-known in France. Translated into French as early as 1543, it continued to be edited both in the original and in translations throughout the seventeenth century.[20] The publication in 1663 of a "Joconde" written in verse by Bouillon prompted La Fontaine to write his own version; a full-blown *querelle littéraire* ensued in which the relative merits of the two poems were hotly debated. The major document in the debate, the *Dissertation sur Joconde*,[21] vehemently defended La Fontaine's poem. The general consensus was that Bouillon's work was a translation of Ariosto, while La Fontaine treated his source as a point of departure for creating his own tale. In an age when "originality" was not necessarily perceived as superior to a faithful imitation of the source, views were divided, "les uns prétendant que le conte était devenu meilleur, par le changement qu'on y a fait; et les autres, au contraire, soutenant qu'il en était tellement défiguré qu'il n'était pas connaissable."[22]

"Joconde" functions well as an introductory tale with its characteristic counterbalancing of the foibles and machinations of both sexes and its focusing on the cognitive framing involved in sexual deceit. In the long 243-line expository phase of the narrative, the desire to see and the desire to know on the part of the king, Astolphe, and of Joconde provide the impetus for their vengeful quest to seduce women (ll. 244-480). When at the end of their adventures the book of their escapades is "plus que plein," their changed perspective about women and marriage brings Joconde and Astolphe to a wisely doubtful framing of their own knowledge. They now understand that *voir* and *savoir* are not necessarily identical.

The two husbands discover first that their wives are unfaithful. The voyeurism often present in the theme of cuckoldry[23] comes to the fore as Joconde *sees* his wife in bed with a valet (l. 95) and then peers through cracks in the wall to see the queen disporting herself with the court dwarf:

> Le Romain, sans beaucoup de peine,
> Les vit en approchant les yeux
> Des fentes que le bois laissait en divers lieux.
> (ll. 178-80)

The king must *see* for himself:

> Je vous tiens, dit Astolphe, homme digne de foi;
> Mais la chose, pour être crue,
> Mérite bien d'être vue:
> Menez-moi donc sur les lieux.
> Cela fut fait, et de ses propres yeux
> Astolphe vit des merveilles.
> (ll. 231-36)

The women have won the first round in the game of deception. Their infidelity, far from evoking a misogynist view of the frivolity of women, shows up the irresponsible characteristics of their husbands:

> Si en effet les femmes, celle de Joconde, celle du roi Astolphe se montrent infidèles, ce n'est certes pas parce que la nature féminine comporte ce trait constitutif, selon les préjugés de la tradition gauloise et misogyne, mais bien parce que leurs maris les ont poussées à bout par leur vanité, leur égoïsme et leur indifférence. (Sweetser "Images féminines," 202)

Neither of the two lovers chosen by the wives to replace the husband is as physically attractive as the husband. One is an ugly valet and the other is the court dwarf. This, however, does not make the substitution more acceptable to the husbands. On the contrary, the very lack of equivalence in the substitution makes the affront to the two husbands all the more unpalatable: underlying the wives' choice is an implicit judgment that the ugly valet and the court dwarf are superior to the husbands.[24] In such a classic comic situation the affront is all the greater because the wife appears to consider her husband less worthy—at least in some aspects—than someone he judges on all accounts to be far beneath himself. Joconde laments:

> Encor si c'était un blondin!
> Je me consolerais d'un si sensible outrage;
> Mais un gros lourdaud de valet!
> C'est à quoi j'ai plus de regret.
> (ll. 114-17)

The discovery of the queen's infidelity replicates Joconde's earlier detection of his own situation:

> Je ne suis pas le seul; et puisque même on quitte
> Un prince si charmant pour un nain contrefait,
> Il ne faut pas que je m'irrite
> D'être quitté pour un valet.
> (ll. 187-90)

Spurred on by their new knowledge, the king and Astolphe plan a revengeful counterattack. The structure of the narrative forming the central part of "Joconde" is of the *trompeur trompé* type, a deceit pattern that juxtaposes the wives' dupery to the husbands' escapades:

> Nos femmes, ce dit-il, nous en ont donné d'une;
> Nous voici lâchement trahis:
> Vengeons-nous-en, et courons le pays.
> (ll. 244-46)

The text alludes summarily to their amorous adventures and to a "livre blanc" (La Fontaine's invention) into which they enter the names of the women they seduce—their trophies of conquest—as so many signposts along the path of their adventure.

Ironically, however, even after so many experiences, Joconde and Astolphe are duped once more by a woman. The culminating and decisive dupery in "Joconde" presents a pattern of embedded deceit, a situation in which the deceivers believe they are still acting as *trompeurs trompés,* whereas, in truth, their supposed victim is playing the same game with them. This final assertion of female dominance in the sexual game prevents their exploits from becoming a simple celebration of male sexual prowess. Joconde and the king arrive at an inn where they decide to seduce the innkeeper's daughter. They consider her as little more than a desirable object: "Ayons quelque objet en commun" (l. 311); a woman is a commercial commodity: "Une grisette est un trésor" (l. 326). They make an exchange pact with her: she consents to sex if they give her a precious ring: "Le marché se conclut" (l. 363).[25] La Fontaine makes a significant change in the *marché* by giving the young woman the power to negotiate for herself; Ariosto's text presents a male-to-male exchange where the two men strike a bargain with her father and then take her off to another village.

Acts of deceit are built upon false or deficient knowledge states. In a *trompeur-trompé* structure the deficient knowledge state assumed by the deceiver to be the lot of the deceivee is really his own. The first mistake of Astolphe and Joconde is in their cognitive framing of the innkeeper's daughter as a *pucelle*:

> Prenons, dit le Romain, la fille de notre hôte;
> Je la tiens pucelle sans faute,
> Et si pucelle, qu'il n'est rien
> De plus puceau que cette belle;
> Sa poupée en sait autant qu'elle.
> (ll. 334-38)

She subverts the exchange not only by accepting the *marché* while cheating on the value of the "goods" she provides, but by managing to spend a night in bed with three men, making love only with her lover, while each of the other two believes his companion is monopolizing the young woman's favors. The deceit is accomplished by a substitution that triggers in the deceivees a false cognitive framing of the situation:

> Et de ceci, ni de cela,
> Ne se douta le moins du monde,
> Ni le roi lombard, ni Joconde.
> Chacun d'eux pourtant s'éveilla,
> Bien étonné de telle aubade.
> (ll. 422-26)

The discovery scene is prolonged (ll. 440-75) as La Fontaine's narrator relishes the cognitive dissonance between the perspectives of Astolphe and Joconde, both of whom, as the reader knows, are in error:

> Comment? dit le Romain: mais vous-même, à propos,
> Vous avez fait tantôt une terrible vie.
> Moi? dit le roi, j'ai toujours attendu.
> (ll. 444-46)

This new deception by a woman furnishes the impetus for their desire to return home to their wives. The tale has been a long process of discovery and reassessment:

> Retournons au logis par le plus court chemin:
> Si nos femmes sont infidèles,
> Consolons-nous, bien d'autres le sont qu'elles.
> (ll. 492-94)

The vanity and egoism of the two men have been transformed into a subversive wisdom; Astolphe begins to understand that the social order itself-and arranged marriages in particular-are to blame for the infidelity of women:

> Peut-être que l'absence, ou bien la jalousie,
> Nous ont rendu leurs coeurs, que l'Hymen nous ôta.
> (ll. 510-11)

Sweetser very perceptively observes that the lesson of "Joconde" is not the castigation of the infidelity of women, but rather an illustration of how men's egoism causes women's infidelity:

> Cette leçon implicite représente un écart notable par rapport à la position traditionnelle du conte gaulois à tendance misogyne que le titre semblait annoncer au départ. De façon indirecte, La Fontaine nous propose des vues modernes sur les obligations réciproques du mariage. (Sweetser, "Images féminines," 204)

But, the text asks, is the exchange of love on an equal basis even conceivable between marriage partners? Although La Fontaine's source, *Orlando furioso*, is highly misogynist, the text of "Joconde" transcends that tradition. Near the conclusion the narrator indulges in a few lines of utopian dreaming about a future where husbands and wives will be faithful:

> La constellation changera quelque jour:
> Un temps viendra que le flambeau d'Amour
> Ne brûlera les coeurs que de pudiques flammes:
> A présent on dirait que quelque astre malin
> Prend plaisir aux bons tours des maris et des femmes.
> (ll. 495-99)

For the present, however, *cocuage* is a fact of life and "epistemological disquiet" an inevitable accompaniment of the marital status.

The importance of the cognitive dimension throughout La Fontaine's *Contes* is signalled by a further significant addition to the source material from Ariosto. The passage in question postulates the existence of a magical false framing of reality. In such a deceitful hypothetical world, comments Astolphe, even the appearance of sexual infidelity may in fact not be what it seems to be. Perhaps the wives of Astolphe and Joconde have been bewitched; or perhaps the husband-observers have themselves stumbled into a magical entrapment and been led to believe things that were not true:

> D'ailleurs tout l'univers est plein
> De maudits enchanteurs, qui des corps et des âmes,
> Font tout ce qu'il leur plaît: savons-nous si ces gens
> (Comme ils sont traîtres et méchants,
> Et toujours ennemis soit de l'un, soit de l'autre)
> N'ont point ensorcelé mon épouse et la vôtre?

Et si par quelque étrange cas,
Nous n'avons point cru voir chose qui n'était pas?
Ainsi que bons bourgeois achevons notre vie,
Chacun près de sa femme, et demeurons-en là.
(ll. 500-9)

As the author of the *Dissertation sur Joconde* claims, Ariosto's text is an exemplary tale showing women as unfaithful creatures and portraying Joconde's tragic experience. La Fontaine's text, on the other hand, attenuates the criticism of women and expands the cognitive dimension of the text.

"Joconde" constitutes a prelude to the *Contes*, where the recurring leitmotif will be the "bons tours des maris et des femmes" (l. 499). The emphasis has been placed squarely on the cognitive framing of reality; the stage is now set for the *Contes* that follow, and the reader is prepared to enter the playful and sometimes make-believe world of cuckoldry and sexual deceit.

NOTES

1. For Philippe de Lajarte the systematic subversion of conventional exchange patterns constitutes the defining element of all *contes facétieux:* "De cette culture (bourgeoise) de l'échange, cependant, le conte facétieux entreprend... la *subversion* systématique" (330). In Mary Jane Stearns Schenk's chapter on "Deception and Misdeed: The Morphology of the Fabliau" (37-64), where she analyzes "devious methods to carry out a hostile or antisocial action against another" (56), deception and the misdeed constitute the primary narrative functions (F5 and F6) of the material that is destined to violate certain social conventions. Schenk's function F4A is communication and includes the transfer of information, as well as the exchange or communication of objects. See also A. D. Mikhailov, "Genèse de la satire antibourgeoise dans les fabliaux."

2. See Roland Barthes, *S/Z* : "Le Récit; monnaie d'échange, objet de contrat, enjeu économique, en un mot *marchandise*" (95). A.-J. Greimas analyzes the distribution, circulation, and manipulation of knowledge in a text as a "cognitive exchange" ("The Cognitive Dimension of Narrative Discourse"); for his semiotic analysis of the transfer of objects in a narrative structure, see "Un Problème de sémiotique narrative: Les Objets de valeur."

3. This approach has become widely accepted; I cite at random the following studies: Eugene Vance, "Chrétien's *Yvain* and the Ideologies of Change and Exchange," *Yale French Studies* 70 (1986): 42-62; Claude E. Bernard, "La

Problématique de l'échange dans *Le Chef-d'oeuvre inconnu* de Balzac," *L'Année balzacienne* 4 (1984): 201-13; Pierre Testud, "Echange et change dans *Le Mariage de Figaro*, ou le cher ruban de Chérubin," *La Licorne* 7 (1983): 41-52; Brigitte Cazelles, "Mots à vendre, corps à prendre et les troubadours d'Aquitaine," *Stanford French Review* 7 (1983): 27-36.

4. See Thomas Hahn, "Money, Sexuality, Wordplay, and Context in *The Shipman's Tale*."

5. Money as a means of seducing a woman appears as a major theme in "A Femme avare galant escroc," "La Coupe enchantée," and "Le Petit Chien qui secoue de l'argent et des pierreries." Men, as well as women, fall prey to avarice—the husband in "Le Magnifique" provides a rendezvous with his wife in exchange for a horse, and the husband in "Le Petit Chien qui secoue de l'argent et des pierreries" prostitutes himself in order to possess a beautiful castle.

6. On the relationship of money and language as modes of communication, see Geneviève Vaughan, "Communication and Exchange." The Introduction to Marc Shell's *The Economy of Literature* (1-10) provides a number of provocative ideas and bibliographical references on language as an economic exchange.

7. *Commerce* designates social relations in a general sense in the tale "On ne s'avise jamais de tout": "Interdisait tout commerce à sa femme" (l. 2); in "Les Quiproquos" (pp. 921-26) the word refers ambiguously to an economic transaction as well as to a sexual act: "Alix d'abord rejette un tel commerce" (l. 82); in "L'Oraison de saint Julien" *commerce* designates an illicit sexual relationship: a marquis is afraid of being discovered, "dans son commerce au logis de la dame" (l. 155). All these meanings are current in the seventeenth century—see Furetière's *Dictionnaire* and Le Robert, *Dictionnaire historique*. In juridical terms the equation of *commercium* with *permutatio*, or the exchange of goods, was current in the sixteenth and seventeenth centuries; because exchange involves the notion of reciprocity, the tendency was to include all contracts under the term *commercium*. See Poughon, *Histoire doctrinale de l'échange*: "*Commercium* et *permutatio* sont deux termes synonymes qui peuvent soit exprimer l'échange au sens strict et juridique du terme, soit traduire la généralité de tout contrat" (36).

8. The number of times *marché*, *marchand*, *marchandise*, and *marchander* appear in their various forms in the *Contes* is high by comparison with occurrences in the *Fables*—of the 65 occurrences in both works, 40 are in the *Contes*.

For the term *commerce*, 5 of 6 occurrences are in the *Contes*; for *échange*, 2 of 7; for *troquer* and its variants, 17 of 18; for *négoce*, 2 of 2. See Tyler's *Concordance*.

9. See Norris J. Lacey, "Fabliaux Women": "Until recently, critics and readers consistently recognized in these works a fundamental misogyny; some chose to ignore the fact, while others apologized for it—but few denied it. While it was customary to point out that a particular woman might be praised in a fabliau, the fact that she was most likely praised for her craftiness turned it into a backhanded compliment at best." See, as well, Roberta L. Krueger, "Love, Honor, and the Exchange of Women in *Yvain*: Some Remarks on the Female Reader"; Lesley Johnson, "Women on Top: Antifeminism in the fabliaux?"; and Marie-Thérèse Lorcin, *Façons de sentir et de penser: les fabliaux français*.

In the case of Boccaccio, La Fontaine's principal source, some reevaluation of his misogyny is also occurring—see M. Olsen who sees even the tales generally considered antifeminist as directed principally against authority: "[L]a ruse se dirige contre l'autorité et pas contre la femme d'abord objet, puis sujet consentant... Elle continue à être valorisée" (90). See also, Charles Franco, "Boccaccio's Woman: A Liberated Woman?" and Tauno Nurmela, "La Misogynie chez Boccace."

10. See John D. Lyons, "Author and Reader in the *Fables*" and Marc Soriano, "Des *Contes* aux *Fables*."

11. In the fabliau one problem in the depiction of women is that sexual appetite can be used as a negative indicator of the base and evil nature of women. In the opinion of Olsen, Boccaccio, to a large extent, manages to distance himself from this tradition by "une prise de distance consciente d'une tradition où la nature basse de la femme était affirmée par la description durative ou itérative de ses amours, par le nombre des amants et par l'omission des débuts de relations" (97-98).

12. Of course the theme of the "mal mariée" was common in medieval literature—see Marie-Thérèse Lorcin, "Le Mariage dans les fabliaux français."

13. In the course of this study I use the term "seduction" to cover all those activities that are ultimately directed toward leading someone to having sexual intercourse. The difficulty with the word is that, like the seducer himself, it conceals under a refined exterior a multitude of disturbing motivations. While in some contexts seduction consists in ploys of flattery or misrepresentation that

lead to a woman's consent, there are other instances in the *Contes* where the "seduction" does not involve consent and is really a case of rape. I would put into the latter category, for example, "Clochette," the story of a young man who lures a shepherdess off into the woods. See Dianne Hunter, especially chapter 8, on the use of the terms "rape" and "seduction."

14. The etymology of the word *cocu* and the origin of the symbol of horns are mysterious and much debated. The most common view is that *cocu* is derived from the cuckoo, a bird that lays its eggs in other birds' nests. There are some who hold that the etymology of the word also derives from a second meaning of *cuculus*, a yellow spring flower—see Pierre Guiraud, *Etymologie des mots obscènes*. As for the horns, for discussion of the various theories, with special emphasis on what is termed "the capon theory, see Robert Bates and Gregory C. Richter, "The Capon Theory of the Cuckold's Horns: Confirmation or Conjecture?"

15. On the double standard and its connection with a patriarchal society in which women are considered the property of men, see Keith Thomas: "The absolute property of the woman's chastity was vested not in the woman herself, but in her parents or her husband. And it might be sold by them" (213).

16. For an application of the sexual exchange theories of Lévi-Strauss and Freud to gender theory, see Gayle Rubin, "The Traffic in Women: Notes on the 'Political Economy' of Sex."

17. Douglas Bruster tends to view cuckoldry as a male-to-male exchange; in my opinion it is more appropriate to construct a model of cuckoldry as a female substitution in La Fontaine's *Contes*. See Bruster's article, "The Horn of Plenty: Cuckoldry and Capital in the Drama of the Age of Shakespeare."

18. In the fabliaux too, the jealous husband's stupidity acts as a foil to the cleverness of his scheming wife—"C'est la sottise du mari qui permet la ruse de la femme. Ainsi le mari jaloux, avare et usurier, mais muni de cet indispensable attribut qu'est le non-savoir, l'incompréhension, la stupidité, est le pendant de l'intelligence rusée" (Brusegan 150).

19. This "epistemological disquiet" is discussed at some length by Katharine Eisaman Maus, "Horns of Dilemma: Jealousy, Gender, and Spectatorship in English Renaissance Drama."

20. See Collinet on translations of *Orlando furioso* and on the popularity of the tale of Joconde, "un classique de la littérature satirique sur l'infidélité des

femmes" (1346-47). Collinet cites incorrectly canto XVIII as the source; it should be canto XXVIII. For early French translations, see A. Cioranescu, *L'Arioste en France des origines à la fin du XVIIIe siècle,* vol. I, 76-103.

21. See Boileau, *Oeuvres complètes*, 309-24. The work has been attributed to Boileau; for a résumé of the controversy still surrounding its authorship, see Beugnot and Zuber, 99-100.

22. See Collinet edition for this quotation from the *Journal des savants*, Jan. 26, 1665, and other details on the controversy surrounding the two versions of "Joconde" (1347-49).

23. Katharine Eisaman Maus develops an analogy between the spectator-spectacle relationship and the erotic male-female attraction. Maus relates the voyeuristic aspects of the theme to the tale of Acteon who, in Ovid's *Metamorphoses*, gazes upon Artemis and her companions as they bathe (568).

24. See Jane Merino-Morais's analysis in semiotic terms of the "écart" between the lovers and the husbands (103-8).

25. See Merino-Morais's study of economic exchange structures and sexual contracts in "Joconde": "Les performances subséquentes de ce sujet-duel peuvent être désignés à leur tour du nom de contrats sexuels. D'un autre point de vue, on pourrait caractériser les accords entre sujets et objets comme des opérations d'échange" (101).

CHAPTER TWO

UNDER THE PEAR TREE:
FRAMES OF DECEIT IN FIVE "MAGICAL" TALES

While magic is a frequently recurring motif in folklore and fairy tales, La Fontaine's *Contes* resemble more closely the fabliau tradition in which magic is far less common. There are, nevertheless, five tales where magic or pseudomagic constitutes the space-builder that falsifies the frame of interpretation and brings about deception. Each of these magical *Contes*— "La Gageure des trois commères," "La Coupe enchantée," "Le Petit Chien qui secoue de l'argent et des pierreries," "Féronde, ou le purgatoire," and "La Jument du compère Pierre"—presents a clear juxtaposition of divergent mental frames at the diegetic level of the text. The clash between reality and illusion is especially pronounced in deceit operated through magic or pseudomagic, which makes this an appropriate way to begin the study of patterns of deceit in La Fontaine's *Contes*.[1] The magical tales implicitly ask the question posed at the end of "Joconde," the first tale of the collection: "Et si par quelque étrange cas/ Nous n'avons point cru voir chose qui n'était pas?" (ll. 506-7).

In "La Gageure des trois commères" La Fontaine merges three stories, two from Boccaccio (VII, 8 and 9) and another from an unknown source, by creating a motivating initial scenario in which three women trade tales about their husbands. The one who recounts the least clever trick will pay a fine; the judgment will be rendered by Macée:

> Le moins bon tour payera quelque amende.
> Nous le voulons, c'est ce que l'on demande,
> Dirent les deux. Il faut faire serment
> Que toutes trois, sans nul déguisement,
> Rapporterons, l'affaire étant passée,

Le cas au vrai; puis pour le jugement
On en croira la commère Macée.
(ll. 35-41)

La Fontaine's narrator does not, however, delegate the narrative voice to the three "commères"; this tale, like the others, is related in the third person and remains firmly controlled by the voice of the heterodiegetic narrator, that is, in Gérard Genette's terms, the narrator who is not part of the storyworld.

The pseudomagical story told by the second woman (ll. 125-215) has many analogues both in European and oriental literature, but there are basically two versions.[2] Chaucer's "Merchant's Tale" (*Canterbury Tales,* 1245-2418) follows the version in which the husband is physically blind, but has his sight restored by Pluto, the king of the fairies—just in time to see his wife cavorting with a young man up in a pear tree. The supernatural power is real and exercized by external agents, Pluto and his wife Proserpina, the queen of fairyland, both of whom are spectators to the scene being played out under the pear tree. Prosperpina's role is to provide the wife with a perfect riposte to the accusation of infidelity. In the many tales that follow this version, the wife answers that this is the way to cure her husband's blindness. In the many tales that follow the second version, like Boccaccio's and La Fontaine's, the husband is not physically blind; he sees what is going on under the pear tree, but is made to believe that it is all an optical illusion wrought by the enchanted tree, or in some tales by a magical window. The reference to a folkloristic belief in primitive magic and talismans is far more pronounced in the tradition followed by La Fontaine. In his version the husband sees, but is made to believe that what he sees is nothing more than an illusion; in Chaucer's version the husband sees and knows what is happening, but his wife provides a false explanation. In La Fontaine's version the deception is accomplished by creating a false and illusory belief in the magical properties of the pear tree, whereas in Chaucer's version the deceptive change in the husband's belief frame comes about through his wife's lie.

As the narrative begins, the wife assumes the role of the deceiver by manipulating circumstances so that she can deceive her husband. Seated under a pear tree with her husband, she orders the valet, in a gesture reminiscent of Eve in the garden of Eden, to fetch her some pears:

La dame dit: Je voudrais bien goûter
De ce fruit-là: Guillot, monte, et secoue
Notre poirier. Guillot monte à l'instant.
(ll. 132-34)

The valet, Guillot, constitutes the extramarital forbidden fruit—the object of her desire and the motivation for her deceit. The "poirier" will become a phallic symbol before the end of the tale. A curtain of illusion falls over the scene as the valet, acting as an accomplice, pretends to see the husband making love with his wife:

> Vraiment, Monsieur, commence-t-il à dire,
> Si vous vouliez Madame caresser,
> Un peu plus loin vous pouviez aller rire,
> Et moi présent du moins vous en passer.
> (ll. 139-42)

The space-builder for shifting the husband's cognitive frame to the doubtful mode and then to a counterfactual framing of the situation is the pretense of the valet and of the wife. Pretense is an imaginative and more dramatic form of lying that induces false belief and acceptance of an alternate world that is deemed by the deceivee to be real. Pretense is the true cognitive spacebuilder in pseudomagical tales, but magic is falsely seen by the deceivee as the operator of deceit. The valet pretends, "fait semblant," to see the husband and wife making love. He pretends to have a counterfactual cognitive frame through which he views the situation. The wife fortifies the valet's claim through her own game of pretense by denigrating the valet as a madman and claiming to know nothing of the deceit, although she is its chief perpetrator:

> La dame dit: Que conte celui-là?
> Je crois qu'il rêve: où prend-il ces nouvelles?
> Qu'entend ce fol avecque ses ébats?
> Descends, descends, mon ami, tu verras.
> (ll. 153-56)

La Fontaine highlights the theatricality of the scene by indicating the speakers' dialogue as if the text were a play. The spell of the illusion has taken over as the genre disguises itself temporarily, pretending to be drama rather than poetic narrative:

> GUILLOT
> Est-ce être fou que de voir ce qu'on voit?

LA FEMME

Et qu'as-tu vu?

GUILLOT

J'ai vu, je le répète,
Vous et Monsieur qui dans ce même endroit
Jouiez tous deux au doux jeu d'amourette,
Si ce poirier n'est peut-être charmé.
(ll. 166-70)

The valet has alleged a perspective, or cognitive frame, up in the pear tree, from which he claims to have seen something the husband knows to be false. The valet knows that the husband will judge such a perception of the scene as counterfactual, although he will not understand what motivates the counterfactual framing. The husband's cognitive frame initially is factual and corresponds to reality. He knows that he and his wife have not been making love. The two contradictory perceptions of the scene, the factual and the counterfactual, are juxtaposed in his mind. The tactics of the wife and her lover in this preliminary stage of the deception eventually, however, bring about the husband's doubtful framing of the situation. A doubtful interpretive frame comes into play when the husband, seeking to corroborate the valet's assertion, climbs the pear tree for the first time. When he sees his wife in the arms of the valet, the doubtful frame shifts quickly back to the factual framing (he believes they are making love and, in fact, they are) because he still trusts the empirical evidence. Enraged at what he sees, he descends rapidly—in fact, just in time, it would seem, to save his honor:

Le maître à peine est sur l'arbre monté,
Que le valet embrasse la maîtresse.
L'époux qui voit comme l'on se caresse
Crie, et descend en grand'hâte aussitôt.
(ll. 174-77)

The preliminary shifting of the husband's interpretation of the situation could be represented as follows, in which the pretense generates a doubtful frame that is subsequently replaced by a factual frame (see Figure 2.1 below).

After he climbs down from the pear tree the stage is set for the main act of deception as the deceitful pretense is resumed. Again the interpretive frame of the husband's mental space is weakened by the protestations of his wife who claims that nothing took place and suggests that the husband was dreaming. With both

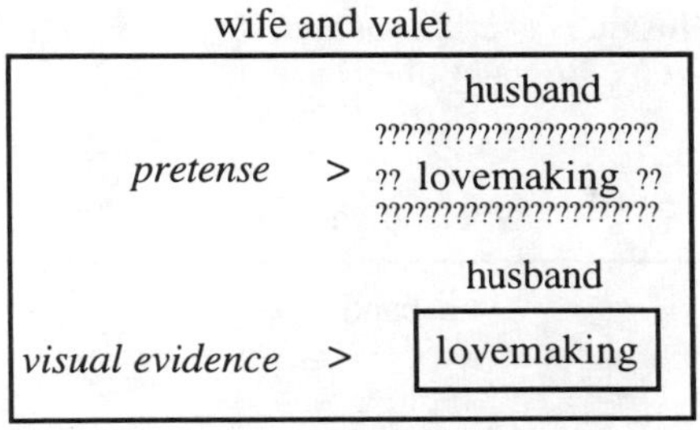

Figure 2.1

the wife and Guillot denying what the husband has seen, the husband, with fewer doubts than the first time, climbs the tree again. While he sees exactly the same scene as before, his interpretive frame shifts so that he now construes the scene of lovemaking beneath him as not factual. This revised frame of belief (he does not believe that what he sees is true) is an illusion. Both the valet and the wife know that the husband's perspective is false. La Fontaine added to Boccaccio's story the second ascent of the husband into the pear tree, providing more cognitive interplay as well as a pleasing parallel structure.

The inversion of the positions of the valet and the husband, from being up in the tree to being below it, as well as the husband's double ascent and descent are figures of the inversion of truth taking place in this tale. What the husband sees is true, but he believes it is false. What the valet claims to see is a fabrication, and he knows it is. The wife and the valet transform the initial imaginary lovemaking scene into a reality when they enact the love scene.

The major act of deceit accomplished by the cognitive space-builder of pseudomagic in this tale appears in the following diagram as a pattern of embedded false frames of belief (see Figure 2.2 below). By their pretense the deceivers have created the husband's false belief that the "magical" pear tree has given him an erroneous perception of the lovemaking scene. Upon the husband's final descent the three characters agree that the only logical conclusion is to blame the pear tree:

> Ne cherchez plus, leur dit-il, d'autres causes;
> C'est ce poirier, il est ensorcelé.
> Puisqu'il fait voir de si vilaines choses,
> Reprit la femme, il faut qu'il soit brûlé.
> Cours au logis; dis qu'on le vienne abattre.

Je ne veux plus que cet arbre maudit
Trompe les gens. Le valet obéit.
(ll. 202-08)

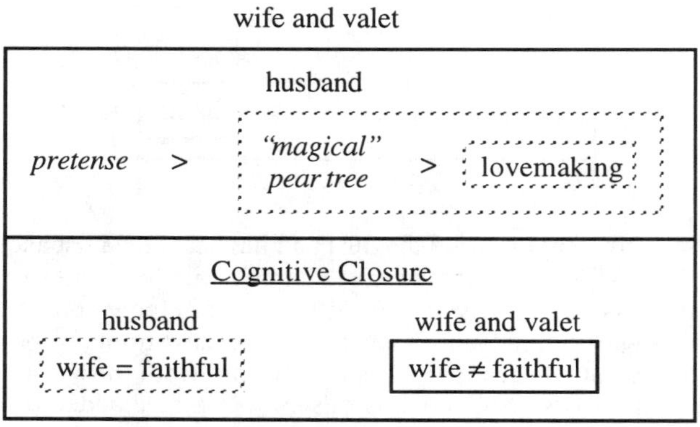

Figure 2.2

The wife's order to cut down the pear tree, a gesture symbolizing castration, brings closure. This destructive action builds up the husband's comforting, but false belief, that order has been restored. There is no final scene of discovery or revelation in the tale, so the cognitive closure is a juxtaposition of the true and the false belief spaces of the principal characters.

The pear tree acquires the erroneous status of being the instrument of false perception, supplanting in the mind of the husband the true perpetrators of the deceit. The tree, like a parody of the mythical tree of Eden, is declared by all three characters to be the source of false knowledge. The valet first suggests the role of the pear tree: "Si ce poirier n'est peut-être charmé" (l. 170). The husband accepts this as a possible explanation just before his second ascent into the tree: "Notre poirier m'abuse assurément" (l. 197). This is confirmed by what he sees: "Ne cherchez plus, leur dit-il, d'autres causes; / C'est ce poirier, il est ensorcelé" (ll. 202-3). The tree is a pseudomagical talisman that makes one believe in the reality of nonexistant things: "Et si par quelque étrange cas / Nous n'avons point cru voir chose qui n'était pas?" ("Joconde," ll. 506-7). Chaucer concludes his tale with a similar reflection:

Beth war, I prey yow; for, by hevene king,
Ful many a man weneth to seen a thing,

And it is al another than it semeth.
He that misconceyveth, he misdemeth.
(ll. 2407-10)

In another tale of enchantment, "La Coupe enchantée," changing the reader's way of looking at marital infidelity—the reader's interpretive framing of this social problem—is from the outset the ostensible goal of the narrator who sets out playfully to prove the paradoxical statement: "Cocuage est un bien." La Fontaine's source for this exemplary tale is Ariosto's *Orlando furioso* (XLII, 69-104; XLIII, 1-49) which he creatively adapts to his own ends.[3] In an extensive prologue (ll. 1-80), of La Fontaine's invention, the *cocu* is presented as an example of someone who lives on illusions, on *songes*, on false mental belief spaces: "Il se maintient cocu, du moins de la pensée, / S'il ne l'est en chair et en os" (ll. 18-19). The narrator goes still further to rectify the common viewpoint on female infidelity, demonstrating with logical aplomb that *cocuage* is unimportant if the husband does not know about it, and relatively insignificant even if he is aware of it:

Quand on l'ignore, ce n'est rien,
Quand on le sait, c'est peu de chose.
Vous croyez cependant que c'est un fort grand cas.
. . . .
Je tire donc ma consequence,
Et dis malgré le peuple, ignorant et brutal,
Cocuage n'est point un mal.
(ll. 24-26, 43-45)

The *cocu*, imagining his wife is unfaithful, lives in a world of false belief spaces that are the irrational product of this eccentric social type:

Ses songes sont toujours que l'on le fait cocu.
Pourvu qu'il songe, c'est l'affaire.
Je ne vous voudrais pas un tel point garantir;
Car pour songer il faut dormir,
Et les jaloux ne dorment guère.
(ll. 7-11)

In the prologue the emphasis on *songe*, the contrast between reality and illusion, between true perspectives and false perspectives, is a preparation for the narrative that follows.

The talisman this time is not a pear tree, but a magical cup that reveals whether one's spouse has been faithful.[4] If a wife has always been faithful, her husband can drink from the cup without spilling its contents. Reminiscent of the interplay of factual and counterfactual perspectives in "La Gageure des trois commères," the narrator's view in "La Coupe enchantée" is that it would be better not to know, to have a false perspective, if one's wife were unfaithful.

The deceit structure in "La Coupe enchantée" is unusual in that Nérie, the primary manipulator, and her goal of seduction are overshadowed in the text as soon as she sets in motion a husband's quest for truth about his wife's fidelity. A sorceress, Nérie, has fallen in love with Damon, a common mortal. Her only way to win him for herself is to deceive him into believing in a falsehood—the infidelity of Caliste, his beloved wife. Nérie first makes him suspicious of his wife's visits from Eraste; his jealousy leads him to create imaginary scenarios, a scene that La Fontaine renders vividly through the *indirect libre* style of discourse:

> Où Damon jusqu'alors avait-il mis ses yeux?
> Car d'amis! moquez-vous c'est une bagatelle.
> En est-il de religieux
> Jusqu'à désemparer alors que la donzelle
> Montre à demi son sein, sort du lit un bras blanc,
> Se tourne, s'inquiète, et regarde un galant
> En cent façons, de qui la moins friponne
> Veut dire: Il y fait bon, l'heure du berger sonne;
> Etes-vous sourd? Damon a dans l'esprit
> Que tout cela s'est fait, du moins qu'il s'est pu faire.
> Sur ce beau fondement le pauvre homme bâtit
> Maint ombrage et mainte chimère.
> (ll. 261-72)

The text here returns momentarily to the motif of *songe* from the prologue, those false perspectives of reality that jealousy generates.

The central part of the narrative action is the carrying out of Nérie's proposal to provide proof of Caliste's infidelity. Nérie uses her magical powers to supply two instruments of deception to Damon: first, by transforming Damon into his suspected rival, Eraste (ll. 273-343), and, secondly, by offering him the magical cup (ll. 344-94).

His metamorphosis is accomplished by a magical means, by rubbing his wrist with "l'eau de la métamorphose" (l. 281). Damon has now effectively moved into the world of falsehood. He has abandoned looking through the doubtful frame of suspicion into a false and fabricated world; he has stepped through the frame and become a "feint personnage" (l. 286). The counterpart of Damon's blurred mental space, based on suspicion inculcated by Nérie, is Nérie's factual mental space, her true vision of Caliste as faithful and of Damon as the dupe of her own manipulative strategies.

Damon's attempt to seduce his own wife under the feigned appearance of Eraste represents for him a means by which to test Caliste's fidelity. He wants to eliminate doubt. By falsifying Caliste's perspective so as to make her believe that he is Eraste, Damon, constructs an embedded deceit pattern and, ironically, undermines his own plan. When Caliste is finally on the point of surrendering to his advances, Damon returns to his true form. Through a magical means Damon, enveloped in doubt, has constructed a factual framing of his wife who falsely believes that she is meeting Eraste.

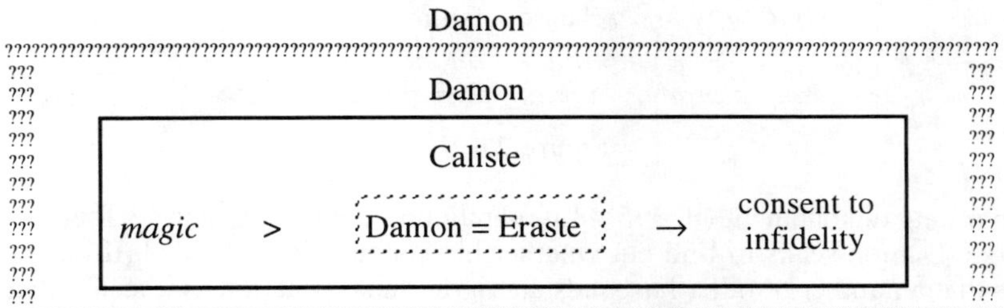

Figure 2.3

His problem, however, is how to reconcile her counterfactual and his factual cognitive frames. Can Caliste be considered unfaithful for being about to succumb to the advances of her husband, even though he was a "feint personnage"? Is he a *cocu*? "L'était-il, ne l'était-il point?" (l. 343). This would be a challenging "cas d'amour" for even the seventeenth-century *précieuses* with their casuistry of love.

Nérie next offers Damon a magical cup that holds out, like the fruit of paradise, a promise to acquire otherwise inaccessible knowledge. The fruit of the pear tree in "La Gageure des trois commères" was only a sham; here the talisman is truly magical. Jealous suspicion has by now distorted Damon's frame of belief to such an extent that he does not really want to believe in his wife's innocence. He continues day after day drinking from the cup until, inevitably, some of the liquid spills:

Ce fut bien là le comble. O science fatale!
Science que Damon eût bien fait d'éviter.
(ll. 376-77)

And yet, despite even this evidence, Damon never fully switches from his doubtful interpretive framing of his wife's infidelity. He understands that the quest for truth through the cup is possibly an exercise in self-delusion, giving rise to more questions than it answers.

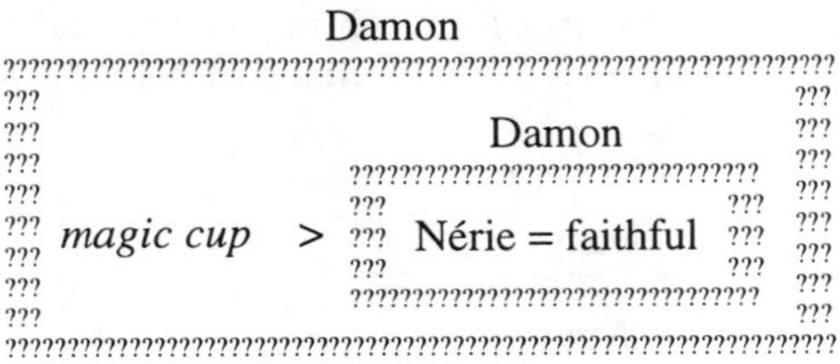

Figure 2.4

In a long dénouement (ll. 395-480), paralleling the prologue in its focus on *cocuage*, Damon seeks to find out whether his jealousy is warranted. If he discovers that a number of other husbands are in the same situation as himself, then he will pardon his wife and be appeased. Damon invites all the male passersby to his table and proposes that they drink from the cup of knowledge: "Voulez-vous savoir si la vôtre \ Vous êtes fidèle?" (ll. 407-8). Almost all of them learn through the magic cup that their wives, too, are unfaithful. At the conclusion, Damon offers the cup to Renaud, Charlemagne's nephew, who wisely refuses to drink. The message of the narrator as it was affirmed in the prologue expresses itself clearly again. What if, Renaud muses, it were due to his own clumsiness, or even the cup's error, and not an infallible magical force, that the cup spilled?

Que sais-je? par hasard si le vin s'épandoit?
Si je ne tenais pas votre vase assez droit?
Je suis quelquefois maladroit:
Si cette coupe enfin me prenait pour un autre?
(ll. 466-69)

Uncertainty, the doubtful framing of his wife's fidelity—the "que sais-je?"—is preferable to the possibility of constructing a counterfactual mental space in which his wife would be unfaithful, and he a *cocu*. He would believe falsely that his wife had been unfaithful—"chose qui n'était pas." It is better not to know and to retain a doubtful faith in one's spouse: "Quand on l'ignore, ce n'est rien" (l. 24), as the narratees were admonished in the prologue. When at the end of this exemplary tale the narrator directly addresses husbands, he proposes Renaud, not Damon as a model; his message is ironic and somewhat cynical:

> Epoux, Renaud vous montre à vivre.
> Pour Damon, gardez de le suivre.
> Peut-être le premier eût eu charge de l'ost;
> Que sait-on? Nul mortel, soit Roland, soit Renaud,
> Du danger de répandre exempt ne se peut croire.
> Charlemagne lui-même aurait eu tort de boire.
> (ll. 481-86)

The enchanted cup, like the pear tree, is an unreliable mental space-builder that provides only a weak basis for building a frame of belief. The philosophical question of the uncertainty of knowledge attained through the senses is again raised, as it was in "La Gageure des trois commères."

La Fontaine emphasizes the metaphysical theme of the fragility of the bases for belief by maintaining throughout his tale of Damon and Caliste an element of doubt about Caliste's infidelity. In Ariosto's *Orlando furioso*, the woman (his three characters are not named), upon discovering that her husband has tricked her into believing he was another man, simply departs to live with the rival. In La Fontaine's tale Caliste is only guilty, after all, of succumbing to the wiles of her own husband. Does that constitute infidelity? Is it better to know, or not to know? And even if the husband thinks he knows, may he not be the victim of deceit, or of an illusion?

Magical illusion is likewise operative in the deceit structures of "Le Petit Chien qui secoue de l'argent et des pierreries" where nothing is as it seems. The source for this *conte* is again Ariosto's *Orlando furioso* (XLIII, 67-144). The forbidden fruit, as in "La Gageure des trois commères," is illicit love. The talisman and instrument of deceit is a magical dog. Anselme, the elderly husband, is forced to revise his view of an unfaithful wife; Argie, for her part, abandons her austere view of the virtuous conduct becoming to a wife. These narrative transformations of the cognitive dimension are accomplished, and to an even greater degree than in "La Coupe enchantée," by altering frames of belief through magical means.

The initial situation is stable: the faithful wife, Argie, at home while her husband is away, refuses the advances of her many suitors including the most attractive one, Atis. Atis, intent on obtaining at least one night with Argie, plays the role of the deceiver. The event that precipitates a change in the situation and provides Atis with the means to seduce Argie is his chance encounter with the supernatural world when he saves a snake's life. Here, the narrator assumes a tone reminiscent of La Fontaine, the fabulist, whose narrator in "L'homme et la couleuvre" (*Fables* X, 1) shows great compassion for the snake. In reply to a man who is about to kill the animal, Atis pleads:

> Ami, reprit Atis, laisse-le; n'est-il pas
> Créature de Dieu comme les autres bêtes?
> Il est à remarquer que notre paladin
> N'avait pas cette horreur commune au genre humain
> Contre la gent reptile et toute son espèce
> (ll. 115-19)

From this moment on, a world of metamorphosis and illusion takes hold of the structure of the tale and determines its outcome. The snake is not a snake at all, but a fairy called Manto who one day a week must take on the form of a snake. She repays Atis by devising a scheme to win over Argie. Manto changes herself into a magical dog called Favori who, among other talents, produces on demand money and precious jewels, while at the same time she gives Atis the appearance of a pilgrim.

The implementation of the initial and most important deception begins when the pilgrim and dog arrive at the country house of Argie. First the *nourrice* is conquered by a demonstration of the dog's capability and by the handsome appearance of Atis. Next, Argie hears from the *nourrice* a series of arguments about the unimportance of her promise to her husband, about the guarantees of secrecy, and above all about the appearance of Atis and his promise to give the dog to her. Infidelity becomes an enticing possibility. Ironically, Argie superimposes on her counterfactual belief in the identity of the pilgrim what she believes is a hypothetical frame: "Si c'était encore le pauvre Atis!" (l. 256):

> . . . Avoir l'effronterie
> De lui mettre en l'esprit une telle infamie!
> Avec qui? Si c'était encore le pauvre Atis!
> Hélas, mes cruautés sont cause de sa perte.
> Il ne me proposa jamais de tels partis.
> (ll. 254-58)

For her the decision will be based on an evaluation "Des merveilles du chien, et des charmes du sire" (l. 290). These are precisely the two misleading factors that act as false space-builders providing an erroneous basis from which Argie finally shifts her view of her role as a faithful wife. The dog's performance in her presence is particularly persuasive:

> Ce n'est pas tout; il se secoue:
> Aussitôt perles de tomber,
> Nourrice de les ramasser,
> Soubrettes de les enfiler,
> Pèlerin de les attacher,
> A de certains bras . . . .
> (ll. 307-12)

And Atis is surprisingly charming and handsome. Later on that night, however, Argie's interpretive framing of the situation shifts radically, and to her great satisfaction, for as soon as she is in Atis's arms, he reveals his true identity. In Ariosto's version Adonio discloses his identity as a final argument just before Argia consents. La Fontaine's revision makes the moment of revelation more dramatic and strengthens the motivation of avarice. The deception of Argie and the cognitive closure of the first half of the tale can be represented as follows:

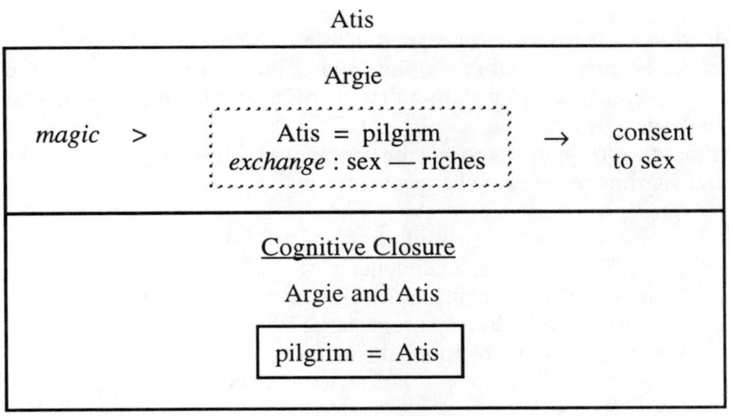

Figure 2.5

Upon his return, Anselme's view of Argie's transgression is initially that of the typical *cocu*:

> La perfide a couvert mon front d'ignominie.
> Pour satisfaction je veux avoir sa vie.
> Poignarde-la; mais prends ton temps.
> (ll. 359-61)

In this second part of the diptych of deception a parallel seduction of the husband takes place as Manto works her magic on Anselme, making him, too, fall into the trap of avarice and shift his view of the situation. She, in the guise of a Moor, carries out this deceit by presenting him with a beautiful palace that will be his on the condition that he takes on the role of an "enfant d'honneur" for two days.[5] Anselme's pact with the Moor makes him, like his wife, a victim of sexual "seduction"; the salacious proposal the Moor makes in lines 421-29 even suggest, as La Fontaine's editors point out, some unprintable words excised from the beginning of line 422:[6]

> Si tu me veux servir deux jours d'enfant d'honneur
> . . . Entends-tu ce langage?
> Et sais-tu quel est cet usage?
> (ll. 421-23)

The allusion to Ganymede and Jupiter's homosexual relationship in the next lines leaves no further doubt as to the Moor's intent. Ariosto evokes the Moor's proposal more discreetly by stating that his offer was identical to the one Adonio made to Argia.

Argie, a secret witness to this entrapment and transformation, feigns not to recognize her husband when she meets him in the palace:

> . . . Est-ce Anselme, dit-elle,
> Que je vois ainsi déguisé?
> Anselme? Il ne se peut; mon oeil s'est abusé.
> Le vertueux Anselme à la sage cervelle
> Me voudrait-il donner une telle leçon?
> (ll. 448-52)

Anselme's own avarice has been exposed as comparable to Argie's. The jealous husband must now put aside his intolerance, his condemnation of her "fort plaisant adultère" (l. 459). The cognitive frame through which he interprets the situation does not alter his judgment of the fact of his wife's infidelity, so much as it shifts his perspective of the infidelity. She explains, juxtaposing his "seduction" with hers: "Du moins n'ai-je pris un More pour galant: / Tout me rend excusable" (ll. 460-61). Anselme and Argie return to their own relationship with a renewed understanding about marital fidelity. The deception of Anselme and the cognitive closure of the narrative is represented in the following diagram:

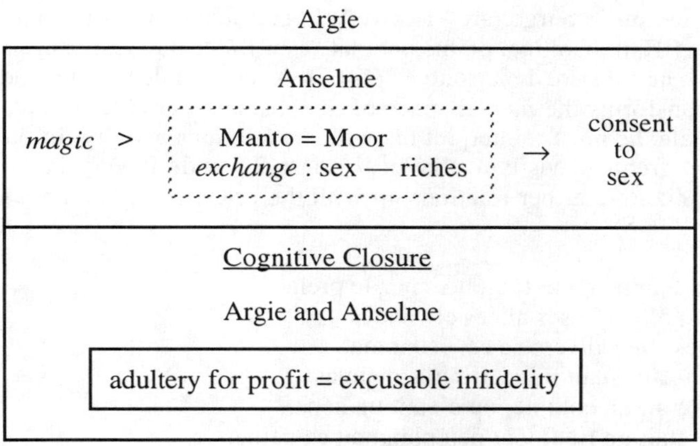

Figure 2.6

The narrator, adding on to Ariosto's version a metanarrative codicil, concludes his text of "Le Petit Chien qui secoue de l'argent et des pierreries" in a proliferation of question marks as the cognitive space is invaded by the inquisitive reader:

> Que devint le palais? dira quelque critique.
> Le palais? que m'importe? il devint ce qu'il put.
> A moi ces questions! suis-je homme qui se pique
> D'être si régulier? Le palais disparut.
> Et le chien? Le chien fit ce que l'amant voulut
> Mais que voulut l'amant? . . . .
>
> (ll. 506-11)

The narrator answers begrudgingly, for the storytelling is already over. The text of the *conte*, however, like "La Coupe enchantée" with its final "Que sait-on?," remains unfinished or "open" to use Umberto Eco's terminology. In "La Coupe enchantée" the narrator urges upon the reader a doubtful cognitive frame: what if your wife's apparent infidelity is not as it appears? In "Le Petit Chien qui secoue de l'argent et des pierreries" the reader's dissatisfied questioning about closure on the diegetic level is only answered by a narrator's exasperated response—an impatient tone that reveals La Fontaine's view that the true interest of the *conte* resides not so much in what happens as in the changing cognitive perspectives engendered by the events.

"Féronde, ou le purgatoire" is a very liberal adaptation of a tale from the *Decameron* (III, 8). Collinet points out that "sauf l'idée du personnage séquestré, presque rien ne subsiste de la source" (1467). A magical deception readjusts and radically transforms the mental space of the husband Féronde so that, as in the preceding tale, he must reinterpret his wife's adulterous activity. Initially suspicious of her frequent visits to the local priest, Féronde is very harsh with his wife, recognizing that her relationship with the priest is not as innocent as she claims:

> Sa femme allait toujours chez le prélat;
> Et prétextait ses allées et venues
> Des soins divers de cet économat.
> Elle alléguait mille affaires menues.
> C'était un compte, ou c'était un achat;
> C'était un rien; tant peu plaignait sa peine.
> (ll. 81-86)

The priest orchestrates an elaborate deceit operation, first sequestering the troublesome husband by making him believe he is in purgatory, and, secondly, bringing him back to life when his wife becomes pregnant. In the first stage of the deception the priest uses a magical "poudre endormante" (l. 111) to put Féronde to sleep so soundly that he does not awaken even during his own funeral. La Fontaine uses the *indirect libre* discourse to advantage when Féronde regains consciousness in a cave that he believes to be his tomb:

> Qu'est-ce cela? songe-t-il? est-il mort?
> Serait-ce point quelque espèce de sort?
> (ll. 120-21)

Through the pretense of the priest's accomplices in the cave, Féronde is made to believe that he is in purgatory and will remain there for a thousand years. Féronde's counterfactual understanding of his situation makes him willing to endure the beatings and fasting imposed on him in the name of purification. So radically does his perspective alter that he resolves not to suspect his wife of anything, if only he can return to earth.

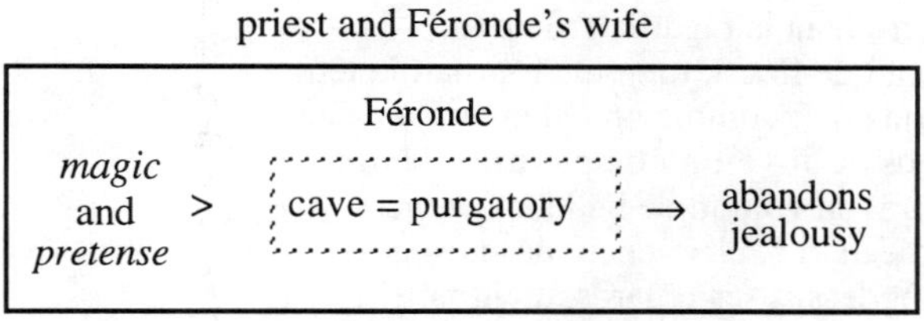

Figure 2.7

In the meantime the priest has been so busy consoling the widow that he is about to become a father in more than name:

> Son soin ne fut longtemps infructueux:
> Pas ne semait en une terre ingrate.
> *Pater abbas* avec juste sujet
> Appréhenda d'être père en effet.
> (ll. 186-89)

The second phase of the deception, the miraculous return of Féronde, prepares the dénouement. To the astonishment of all, the abbé prays so fervently, that Féronde returns from purgatory still firmly believing that he had been dead. He accepts the child as his own, thus adopting a permanently false view of the situation; for Féronde the moment of revelation never takes place. The villagers interpret the event through a counterfactual frame:

> . . . Un si merveilleux cas
> Surprit les gens. Beaucoup ne voulaient croire
> Ce qu'ils voyaient. L'abbé passa pour saint.
> (ll. 195-97)

"La Jument du compère Pierre," like the tale of the pear tree, portrays a naive husband, in this instance Pierre, who is deceived into believing in pseudomagic. Boccaccio's rendering of this tale (*Decameron* IX, 10) is more crude and less detailed than La Fontaine's. At the beginning of Boccaccio's tale the priest's desire for Magdeleine, Pierre's wife, is not described; La Fontaine, however, relishes all the details of the depiction, even inventing as an imaginative depiction of his lust the simile of a dog, complete with a wagging tail (the *queue* motif returns in the climactic scene):

Messire Jean la regardait toujours
Du coin de l'oeil, toujours tournait la tête
De son côté; comme un chien qui fait fête
Aux os qu'il voit n'être par trop chétifs;
Que s'il en voit un de belle apparence,
Non décharné, plein encor de substance,
Il tient dessus ses regards attentifs:
Il s'inquiète, il trépigne, il remue
Oreille et queue; il a toujours la vue
Dessus cet os, et le ronge des yeux
Vingt fois devant que son palais s'en sente.
Messire Jean tout ainsi se tourmente
A cet objet pour lui délicieux.
(ll. 35-47)

In order to satisfy his desires, the priest devises a "plaisant stratagème" destined to overcome the resistance of Magdeleine and to manipulate her stupid husband. He proposes to work a charm so that the peasant's wife will "par art d'enchantement" (l. 78) be a mare during the day and a woman at night. Transportation of the farmer's produce to the market will be greatly expedited and the mare will need to eat only a small amount of grass daily. Lured by the pretense of a "magical" means to attain future prosperity, Pierre is duped into believing in the fabricated cognitive frame; he even encourages his reluctant wife, Magdeleine, to disrobe in front of Messire Jean. Pierre, in fact, removes her last garments. Boccaccio stated matter-of-factly that she took off her clothes, but La Fontaine dwells on the sensuality of the scene—which is heightened by Magdeleine's *pudeur* and hesitations:

Dégrafez-moi cet atour des dimanches;
Fort bien: ôtez ce corset et ces manches;
Encore mieux: défaites ce jupon;
Très bien cela. Quand vint à la chemise,
La pauvre épouse eut en quelque façon
De la pudeur. Etre nue ainsi mise
Aux yeux des gens! Magdeleine aimait mieux

Demeurer femme, et jurait ses grands deux
De ne souffrir une telle vergogne.
(ll. 110-18)

Pierre, like the husband under the pear tree, witnesses Messire Jean making love to his wife, but does not believe what he sees. The deceit is carried out by first duping Pierre into building a hypothetical mental space within which he foresees a prosperous future; the second step in the deceit structure is to shift Pierre's cognitive frame into the counterfactual mode, so that he does not interpret what he sees as he normally would. The curé, claiming to operate his magical transformation of her into a *jument*, touches her all over so that each part of her body comes under the spell. Again La Fontaine indulges in details that were not in Boccaccio's text; the female body is described in geographical terms with a reticence worthy of a seventeenth-century *précieux* poet:

Messire Jean par le nombril commence,
Pose dessus une main, en disant:
Que ceci soit beau poitrail de jument.
Puis cette main dans le pays s'avance.
L'autre s'en va transformer ces deux monts
Qu'en nos climats les gens nomment tétons;
Car quant à ceux qui sur l'autre hémisphère
Sont étendus, plus vastes en leur tour,
Par révérence on ne les nomme guère.
(ll. 133-41)

Pierre, seeing no progress being made, even enters, through his cognitively deficient frame, into the illusory world further and prays for the metamorphosis:

Tant de façons mettaient Pierre en chagrin;
Et ne voyant nul progrès à la chose,
Il priait Dieu pour la métamorphose.
(ll. 146-48)

The supremely magical moment during which Magdeleine's actual transformation into a mare is to be accomplished consists in the act of giving the animal a tail: "car de l'enchantement / Toute la force et l'accomplissement / Gisait à mettre une queue à la bête" (ll. 149-51). Only when it is too late does Pierre see what is really happening:

Jean ne voulant un tel point oublier
L'attache donc: lors Pierre de crier,
Si haut qu'on l'eût entendu d'une lieue:
Messire Jean, je n'y veux point de queue:
Vous l'attachez trop bas, Messire Jean.
(ll. 153-57)

In an ironic closure, wholly of La Fontaine's invention, both the curé and Magdeleine berate Pierre and verbally transform him into a donkey. Because Pierre refused to accept Magdeleine's supposed metamorphosis into an animal, she and Messire Jean castigate Pierre for his *bêtise*:

Le curé Jean, qui lui dit: Foin de toi:
T'avais-je pas recommandé, gros âne,
De ne rien dire, et de demeurer coi?
(ll. 164-66)

Magdeleine is angry, and, very subtly, her innuendoes about Pierre's virility come through the lines by way of the rhyme "malheureux" / "gueux" that echoes the earlier "queue" rhyme (ll. 155-56):

. . . Malheureux,
Tu ne seras qu'un misérable gueux
Toute ta vie: et puis viens-t'en me braire.
(ll. 170-72)

Does the deceivee discover the deceit? In any case, the duped husband rejects any more "magical" machinations of Messire Jean and regains control of the situation. When Magdeleine, again alluding to her husband in animal metaphors ("oison," "veau"), invites the priest to come and transform her into a "jument" every day, Pierre, unlike Féronde or the husband under the pear tree, clearly refuses further commerce with Messire Jean: "Plus de jument, ma mie" (l.183).

In the five magical tales deceptive tactics are deployed in order to create belief frames through which the characters in the tales see or comprehend situations in a way that does not correspond to the factual situation. La Fontaine skillfully constructs and deconstructs belief spaces within the narrative, juxtaposing and superimposing divergent cognitive frames on the same set of circumstances. In "La Gageure des trois commères" the husband's framing of the actual scene of lovemaking between his wife and the valet is juxtaposed with the valet's pretended framing of the scene as involving no sexual activity. The deception is complete when one frame is superimposed on the other in the husband's mental space, so that he sees the lovemaking but does not believe it is taking place. In these tales illusions created by false perspectives are constantly being played with on the diegetic level of the text as they are created through various forms of deceit and cast aside through revelation and discovery.

If within the fictional world, the deceiver, and sometimes one or more accomplices, truly understand the situation and are aware of the false framing activity of those who are victims of the deceit, the reader occupies an even more privileged position. The continually shifting changes in cognitive framing within the diegetic world provide a pleasure for the reader that derives from occupying a superior cognitive space from which to view the multiple perspectives simultaneously, assess their relative validity, and enjoy the ironic double vision of contradictory cognitive spaces. For instance, in "Le Petit Chien qui secoue de l'argent et des pierreries" the reader relishes the dramatic irony of the situation when Argie wishes that she were about to make love to Atis; in fact, that is precisely what she is about to do. The narrator of the *Contes* creates a pact of complicity with the reader by sharing enough information so that the reader's interpretive frame is large enough to comprehend the shifting perspectives of the characters.

Common patterns of deceit and trickery run through all the *Contes*, but the dominant, albeit apparently playful, concern is the one highlighted in the magical tales, that is, the hypothesis that society, the world we live in, and our knowledge of it, is possibly all illusory. We humans are all sitting under the pear tree, thinking we know reality, but what if our perspective is false: "Et si par quelque étrange cas / Nous n'avons point cru voir chose qui n'était pas?" ("Joconde," ll. 506-7). This is the recurring subliminal question put to the reader throughout the *Contes*.

NOTES

1. Of course, magic and pseudomagic are terms that must be understood within the context of the seventeenth-century belief system. What was magical at that time may not be perceived as such in the twentieth century. In the Stith Thompson *Motif-Index of Folk Literature* many feats appear under the heading of magic that, thanks to modern technology and medicine, are no longer considered magical—severed limbs replaced (D2161.5.2.4), transportation by submarine (D1525), or transportation by airships (D1118).

2. For a full discussion of the many versions of this tale, see Germaine Dempster's presentation of the "Merchant's Tale" and her reproduction of several

versions in Bryan and Dempster's *Sources and Analogues of Chaucer's* "Canterbury Tales," 333-56.

3. La Fontaine suppresses the lengthy descriptions of the castle and the women. He adds his own prologue on *cocuage* and inverts the order of the narrative so that drinking from the cup follows Damon's duping his wife under the guise of another man.

4. The folkloric antecedents for the motif of a test determining whether adultery has occurred are numerous—see Stith Thompson, *Chastity tests* H400. The closest to Ariosto's magic cup is H411.4.2: "Magic cup as chastity test: sticks to hands of adulterer." In H411.4: "Magic drinking horn (cup) as chastity test," an unchaste woman cannot drink from the cup.

5. In Pierre Guiraud's *Dictionnaire érotique*, La Fontaine's verse is cited as the example for the entry, "enfant d'honneur-giton."

6. In Hoole's nineteenth-century translation of this passage into English he changes the Moor into a "female Ethiop," explaining: "A liberty is here taken with the original, in varying the circumstances of the story, too gross to admit of the most distant allusion" (*The Works of the English Poets,* vol. 21, 350).

CHAPTER THREE

DISGUISE:
DECEIT BY SEMIOSIS

Disguise generally functions in the *Contes* as a mediated action that advances the plot of the tale. The deceiver constructs an alter ego—a mediator—who acts in his or her place. What a seducer cannot do to approach the woman he desires, can be accomplished through a disguise. Adopting the external appearance of another person is equivalent to delegating the task of accomplishing the desired action to a mediator. As Georges Forestier observes, however, in the case of disguise "[c]e médiateur, c'est lui-même, mais sous une autre identité: un autre lui-même" (408).

Forestier has analyzed in admirable detail the aesthetic functioning of disguise in 404 plays of the period 1550-1680. In some respects his study, *Esthétique de l'identité dans le théâtre français*, is a useful *point de repère* for approaching La Fontaine's handling of disguise as a manipulative strategy in the *Contes*. Forestier understands disguise in a very broad sense, however, including under that rubric instances of mistaken identity and lies about a person's identity:

> Le seul critère en jeu est le critère minimal de l'identité. Que l'apparence soit affectée ou non, que le personnage déguisé soit conscient d'être autre que ce qu'il paraît ou qu'il soit le premier trompé par son identité fictive, on parlera de déguisement dans tous les cas d'interruption momentanée de l'identité relayée par l'imposition d'une identité usurpé. (11)

For the purpose of analyzing deception in the *Contes* it is more productive, in my view, to limit the term "disguise" to a narrower definition, a consciously devised transformation of a character's physical appearance undertaken with the purpose of deceiving.

Disguise, then, is a means of impersonation in which someone deliberately reduces his or her identifying characteristics and takes on the misleading appearance of another person. By substituting the outer appearance of someone else for one's own, the agent of disguise falsifies the perception of the intended victim. Disguise is a substitution involving a semiotic transfer of certain external signifiers that indicate the identity of a person. The victim of the dupery—a naive young woman, or a jealous husband—acquires a false frame of interpretation that permits the agent of the deception to have access to the person who is sought, usually as an object of sexual desire.[1]

In the seven tales that feature disguise, the patterns of deception generated range from simple to complex forms. There are basically two models. In the first, the victim's cognitive frame is falsified so that she accepts the overtures of the seducer because she believes he is someone else. For the seducer, disguise becomes the means of fulfilling sexual desire; for the woman, the process is both an aggression and a legitimization of an otherwise forbidden sexual encounter. There are three of this type, "Le Muletier," "La Mandragore," and "Le Fleuve Scamandre." In the other four ("Le Cocu battu et content," the first tale in "La Gageure des trois commères," "Les Lunettes," "Le Mari Confesseur") the disguise is principally a means of concealing a situation, rather than being a direct tool of seduction.

In "Le Muletier," a tale derived from the *Decameron* (III, 2) of Boccaccio, the disguise is not elaborate but has a dual formulation, one phase functioning to seduce a woman and the other to conceal the deceiver's identity from the husband. In the first disguise, a mule driver intent on sleeping with the queen, Teudelingue, pretends to be the king, Agiluf, by presenting himself in the semi-darkness at her bedroom door. He imitates the king's custom of wearing very little—"presque en chemise"—and carrying a candle. Part of the mule driver's preparation—so that his distinctive smell will not give him away when he gets into the queen's bed—was, in Boccaccio's version, to take a hot bath; La Fontaine's lover is French, he puts on perfume (l. 69)!

Preserving the illusion that he is the king, Boccaccio's mule driver manages to make his silence credible by being a vigorous lover ("piu volte carnalmente la reina conobbe") because he knows that the king, when he is troubled ("mostrandosi turbato") does not speak (Boccaccio 174).[2] La Fontaine, on the other hand, explains the supposed king's unwonted vigor simply by his being a mule driver and not a king. He, in fact, constructs a lengthy passage of 13 lines, presenting the queen's wonderment and the narrator's sociological commentary:

> Notre amoureux fournit plus d'une traite:
> Un muletier à ce jeu vaut trois rois;

Dont Teudelingue entra par plusieurs fois
En pensement, et crut que la colère
Rendait le prince outre son ordinaire
Plein de transport et qu'il n'y songeait pas.
En ses présents le Ciel est toujours juste:
Il ne départ à gens de tous états
Mêmes talents. Un empereur auguste
A les vertus propres pour commander;
Un avocat sait les points décider;
Au jeu d'Amour le muletier fait rage:
Chacun son fait; nul n'a tout en partage.
(ll.76-88)

The initial deceivee, Teudelingue, does not discover the deception. Relegated to a secondary role by the end of the tale, she remains blissfully wrapped up in her counterfactual universe, believing that the *muletier* is Agiluf. The moment of discovery of the first deception comes for Agiluf when his naive wife expresses surprise at seeing him return to her bed so soon—or so she thinks—after he supposedly had left:

Certes monsieur, je sais bien, lui dit-elle,
Que vous avez pour moi beaucoup de zèle;
Mais de ce lieu vous ne faites encor
Que de sortir: même outre l'ordinaire
En avez pris et beaucoup plus qu'assez.
Pour Dieu, monsieur, je vous prie, avisez
Que ne soit trop; votre santé m'est chère.
(ll. 94-100)

In the second phase of deception, the cognitive structure exhibits greater complexity in a reverse disguise that the "deceived" husband provokes in the course of his attempted revenge. Guessing that one of the mule drivers is the guilty party, the king goes to where they are sleeping and feels the pulse of each until he comes to one whose heart is beating fast:

Le coeur battait, et le pouls tout ensemble.
Ne sachant pas où devait aboutir
Tout ce mystère, il feignait de dormir.
Mais quel sommeil! Le roi, pendant qu'il tremble,
En certain coin va prendre des ciseaux

Dont on coupait le crin à ses chevaux.
Faisons, dit-il, au galant une marque,
Pour le pouvoir demain connaître mieux.
(ll. 116-23)

The king's gesture of cutting a lock of the mule driver's hair, so as to identify him the next morning in the daylight, is, however, a plan that is thwarted by the clever *muletier*. He nullifies the sign of his identity as a transgressor by promptly cutting a lock of hair from each of his sleeping companions. The mule driver's stratagem is a reverse disguise in that it makes the others look like him; instead of adopting a misleading identifying mark, he replicates the distinguishing sign. A disguise usually displaces semiotic signs by having the deceiver misappropriate distinguishing features of clothing or other elements of external appearance. Here the semiotic marker—the missing lock of hair—is displaced from the deceiver to a group of others who are drawn into the deception as unconscious collaborators. The king *knows* he has been deceived, but he *does not know* by whom. He discovers the initial deception, but ends up enmeshed in a doubtful frame.

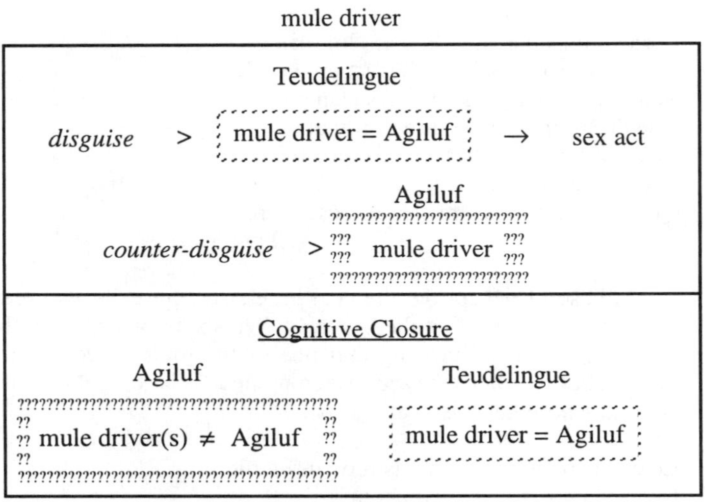

Figure 3.1

The above diagram illustrates, to begin with, how the initial disguise operated by the *muletier* establishes Teudelingue's counterfactual belief space that remains static until the end of the tale. Juxtaposed to the first act of deceit is the counter-disguise, which is also the work of the *muletier* and creates a not entirely factual framing of the situation by Agiluf. Although he knows that he and his wife have been deceived, his interpretative frame is a doubtful, questioning perspective on the identity of the *muletier*. In Boccaccio's text cognitive closure is effected by the king's statement in which he marvels at the cleverness of the mule driver, saying to himself: "Costui il quale io va cercando, quantunque di bassa condizion sia, assai ben mostra d'essere d'alto senno" (175). ["The man I'm looking for may be the basest varlet, but clearly he's no simpleton" (181).] La Fontaine's text omits this passage, however, concluding on the cognitive level with a juxtaposition of Teudelingue's and Agiluf's belief spaces: the queen remains in a state of incomprehension, while the king's perplexity at seeing all the *muletiers* missing a lock of hair expresses itself through an interior monologue:

> Qu'est ceci donc! qui croirait que ma femme
> Aurait été si vaillante au déduit?[3]
> Quoi Teudelingue a-t-elle cette nuit
> Fourni d'ébat à plus de quinze ou seize?
> Autant en vit vers le front de tondus.
> Or bien, dit-il, qui l'a fait si se taise:
> Au demeurant qu'il n'y retourne plus.
> (ll. 135-41)

The pattern of deception in "La Mandragore," a tale loosely based on Machiavelli's play, *Mandragola*, is distinctive in that the would-be lover, Callimaque, manages to enlist as unconscious accomplices in his deception both the husband, Nicia Calfucci, and his wife, Lucrèce, who are the deceivees. The preparatory phase of the deception is elaborate because the accomplices have to become motivated. In order to prepare the circumstances for the disguise to operate, Callimaque, a young man who has recently completed his studies in France, provides some preliminary false information. His lie establishes for both the husband and wife an encompassing false belief frame within which the sexual deception occurs. Nicia Calfucci—more often referred to as Nice in the text—is typecast as a "sot" from the beginning, a man obsessed with his inability to have children:

> Il consultait matrones, charlatans,
> Diseurs de mots, experts sur cette affaire:

Le tout en vain: car il ne put tant faire
Que d'être pêre. Il était buté là.
(ll. 18-21)

Callimaque's ruse begins with convincing Nicia (ll. 64-184) that the root of the mandragore is the best remedy for infertility; this is not a lie, but a folkloric belief of the period. The lie consists in Callimaque's further embroidery on the powers of the wondrous plant: the first one to sleep with Lucrèce after she consumes it may die on the spot:

Ce jus doué de vertu tant insigne
Porte d'ailleurs qualité très maligne.
Presque toujours il se trouve fatal
A celui-là qui le premier caresse
La patiente; et souvent on en meurt.
(ll. 103-7)

Nicia responds skeptically, as Callimaque had foreseen:

Plus de votre herbe; et laissons là Lucrèce
Telle qu'elle est: bien grand merci du soin.
Que servira moi mort si je suis père?
(ll. 109-11)

Callimaque kindly offers (diplomatically using the inclusive *nous*) to find a suitable young man, someone well below the social status of the husband:

Il nous faudra choisir quelque jeune homme
D'entre le peuple; un pauvre malheureux
Qui vous précède au combat amoureux;
Tente la voie, attire et prenne en somme
Tout le venin . . . .
(ll. 124-28)

And yet, this person must not be so uncouth and rough that he will displease the wife, Madame Lucrèce. The husband has to be assured that he will not be accused of murder; the wife has to be relieved of her moral scruples (ll.185-209). The preliminary false cognitive frames of both husband and wife are now in place; they are ready to become collaborators in the deception, but are not fully informed.

Disguise: Deceit by Semiosis

The deceiver's goal in the dupery, the seduction of Lucrèce, is accomplished by means of a disguise that does not so much impersonate someone else as hide the identity of its wearer. Callimaque adopts the guise of a miller, blanketing himself in flour, putting on a blindfold, and adding a false beard:

> Le lendemain notre amant se déguise,
> Et s'enfarine en vrai garçon meunier;
> Un faux menton, barbe d'étrange guise;
> Mieux ne pouvait se métamorphoser.
>
> (ll. 210-13)

He succeeds first of all in duping the husband whose counterfactual interpretive framing of the ensuing action depends on his false belief that the miller is not Callimaque:

> Sur le minuit le mène à messer Nice;
> Les yeux bandés; le poil teint; et si bien
> Que notre époux ne reconnut en rien
> Le compagnon. . . .
>
> (ll. 218-21)

The climax of "La Mandragore" and the scene that is at the narrative heart of La Fontaine's tale is the seduction of Lucrèce in her bed. La Fontaine dwells on this scene (ll. 221-92) highlighting the subjectivity of Lucrèce, which was not done in the original play by Machiavelli where Lucrèce does not even appear on the scene after Act Three. In the *Mandragola* the bedroom scene is narrated by the husband in Act Five—from a truncated and ironic perspective. La Fontaine's text presents directly in a dramatic mode the hesitations, the doubts, and the eventual pleasure of the heroine. The strong presence of the narrator adds to the intensity of the scene as he probes the feelings of Lucrèce ("A ce garçon la perle des Lucrèces / Prendrait du goût?" [ll. 251-52]) and her secret struggles:

> Lucrèce avait jusque-là résisté;
> Non par défaut de bonne volonté;
> Ni que l'amant ne plût fort à la belle:
> Mais la pudeur et la simplicité
> L'avaient rendue ingrate en dépit d'elle.
>
> (ll. 265-69)

Lucrèce is not as easily fooled as her husband. At the beginning of the love scene her suspicions are aroused immediately by a tactile signifier:

> . . . la galande sentit
> Auprès de soi la peau d'un honnête homme.
> Et ne croyez qu'on employât au somme
> De tels moments. Elle disait tout bas:
> Qu'est ceci donc? Ce compagnon n'est pas
> Tel que j'ai cru: le drôle a la peau fine.
> (ll. 236-41)

Callimaque had foreseen Lucrèce's immediate consent, not the unexpected resistance he meets, nor the necessity to drop the disguise in order to achieve his goal. Faced with Lucrèce's resistance (after all, bringing death upon an "honnête homme" is a serious consideration), Callimaque reveals his identity, and so brings about a collapse of her doubtful belief frame as recognition takes place in the very midst of the deception scene:

> . . . Pardon (dit-il) madame.
> Ne vous fâchez du tour qu'on vous a fait;
> C'est Callimaque: approuvez son martyre.
> Vous ne sauriez ce coup vous en dédire.
> Votre rigueur n'est plus d'aucun effet.
> S'il est fatal toutefois que j'expire,
> J'en suis content: vous avez dans vos mains
> Un moyen sûr de me priver de vie;
> Et le plaisir bien mieux qu'aucuns venins
> M'achèvera; tout le reste est folie.
> (ll. 255-64)

The miller is Callimaque! Her understanding of the situation shifts immediately to an acknowledgment that in fact she has been making love with someone she knows. Her complex and culpabilizing reactions translate into an interior monologue in La Fontaine's text:

> Après cela? qu'en pourra-t-il penser?
> Dit-elle en soi, et qu'est-ce qu'il lui semble?
> J'ai bien manqué de courage et d'esprit.
> (ll. 274-76)

Lucrèce quickly overcomes her scruples, however, and enjoys her predicament. The narrator expresses in patriarchal and militaristic terms the triumph of Callimaque:

> Elle y voulut tenir bon, mais en vain.
> Ne lui restant que ce peu de terrain,
> La place fut incontinent rendue.
> Le vainqueur l'eut à sa discrétion;
> Il en usa selon sa passion.
> (ll. 281-85)

In the *Contes* there are a number of deceivees who are pleasantly duped; Lucrèce belongs in this category. The narrator comments through his male perspective: "Heureux sont ceux qu'on trompe à leur profit!" (l. 288). Callimaque still keeps her partially in a counterfactual framing of the situation, as she is unaware that the mandragore has no power whatsoever, and that he is in no danger. The manipulation of Lucrèce, albeit with some degree of consent, is problematic; the text's fixation on Lucrèce's most intimate cognitive space and on its violation verges on perversion.

Nicia Calfucci, on the other hand, never discovers that he was deceived by the disguise and "deceived" by both Callimaque and his wife. He joins the troupe of unaware *cocus*, naively narrating to Callimaque himself what happened that night between his wife and the miller. Closure by means of an ironic vignette adds the final fillip to "La Mandragore." Nicia states confidently that he had assured Lucrèce and the miller that he would know everything about the planned sexual rendezvous and that he could never be duped:

> Et ne pensez, ce lui dis-je, Lucrèce,
> Ni l'un ni l'autre en ceci me tromper;
> Je saurai tout; Nice se peut vanter
> D'être homme à qui l'on n'en donne à garder.
> (ll. 309-12)

The functioning of disguise as a cognitive space-builder in the deception could be schematized as follows. The space-builder in each phase of the tale is put into operation by Callimaque. His preliminary lie about the mandragore covertly subtends the whole of the deceit structure; his credibility depends on it throughout. The frames are juxtaposed and interdependent; the first deception by disguise prepares for the second.

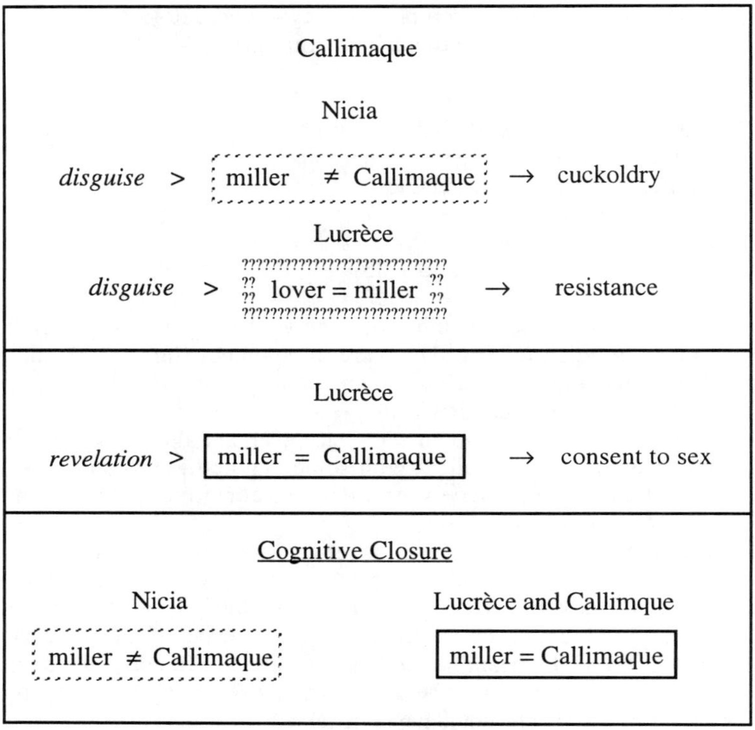

Figure 3.2

If the preparatory phase of the deception was prolonged in the previous tale, the discovery stage is emphasized in "Le Fleuve Scamandre," a *conte* with lyrical overtones that La Fontaine develops from a few narrative elements contained in a letter attributed to Eschine, a Greek orator.[4] The letter tells how the day before their marriage it was customary for young girls to bathe in the river and offer to the god Scamandre their virginity. La Fontaine takes away all the ritualistic aspects, naturalizing the situation so that a young woman one day, because it is so hot, decides to bathe in the river. In the letter, the deception was premeditated: "Notre honnête homme de Cimon se cache dans les herbes du Scamandre, et se couronne de roseaux: c'était un piège tendu pour la circonstance, un tour qu'il avait médité de jouer à Callirrhoé" (Régnier ed. 361). La Fontaine makes Cimon less of a conniver; Cimon just happens to be on the riverbank when the woman appears.

The episode takes place near the ruins of Troy. In the prologue the very mention of Ilion triggers in La Fontaine's classical imagination a lyric and nostalgic response that can only express itself in alexandrines:

> Ilion, ton nom seul a des charmes pour moi;
> Lieu fécond en sujets propres à notre emploi;
> Ne verrai-je jamais rien de toi, ni la place
> De ces murs élevés et détruits par les dieux,
> Ni ces champs où couraient la fureur et l'audace,
> Ni des temps fabuleux enfin la moindre trace,
> Qui pût me présenter l'image de ces lieux?
> (ll. 26-32)

The rhythm of the alexandrine dominates "Le Fleuve Scamandre," punctuated from time to time with octosyllabic verse. As a metatextual replication of the diegetic seduction, the depiction of the heroine (not present in the source) seduces the reader with its obsessively repetitious labial patterns on the phonetic level and its semantic focus on *fraîcheur*, *légèreté*, and *jeunesse*:

> Une jeune ingénue en ce lieu se vient rendre,
> Et goûter la fraîcheur sur ces bords toujours verts.
> Son voile au gré des vents va flottant dans les airs;
> Sa parure est sans art; elle a l'air de bergère,
> Une beauté naïve, une taille légère.
> (ll. 36-40)

As she descends into the water, the gaze of Cimon accompanies her:

> Un antre était auprès: l'innocente pucelle
> Sans soupçon y descend, aussi simple que belle.
> Le chaud, la solitude, et quelque dieu malin
> L'invitèrent d'abord à prendre un demi-bain.
> Notre banni se cache; il contemple, il admire,
> Il ne sait quels charmes élire;
> Il dévore des yeux et du coeur cent beautés.
> (ll. 43-49)

Her foot is highlighted as it dips into the river, already offering a provocative metonymic foreshadowing of the imminent union with the self-proclaimed river-god:

> La belle enfin découvre un pied dont la blancheur
> Aurait fait honte à Galatée,
> Puis le plonge en l'onde argentée,
> Et regarde ses lis, non sans quelque pudeur.
> (ll. 57-60)

Disguising himself with dripping river plants—"Se couronne de joncs, et d'herbes dégouttantes"—Cimon accosts her, presenting himself as Scamandre and eloquently promising an enchanted future:[5]

> Je suis, dit-il, le dieu qui commande à cette onde;
> Soyez-en la déesse, et régnez avec moi.
> Peu de fleuves pourraient dans leur grotte profonde
> Partager avec vous un aussi digne emploi:
> Mon cristal est très pur, mon coeur l'est davantage;
> Je couvrirai pour vous de fleurs tout ce rivage,
> Trop heureux si vos pas le daignent honorer,
> Et qu'au fond de mes eaux vous daigniez vous mirer.
> (ll. 64-71)

Duped by the disguise, the young woman had already entered into a counterfactual belief space in which she is now ready to be persuaded to become a river goddess by succumbing to him. The sexual encounter is discretely suggested—"Conclurent tout en peu de temps"—in keeping with the *bienséances* demanded by an imitation of the heroic style.

The moment of disclosure occurs when the woman points Cimon out to the assembled village people some months later as "le fleuve Scamandre." Unlike what happened in the two previous tales this deceiver is punished, but not seriously:

> On en rit; car que faire? Aucuns à coups de pierre
> Poursuivirent le dieu, qui s'enfuit à grand'erre;
> D'autres rirent sans plus. . . .
> (ll. 99-101)

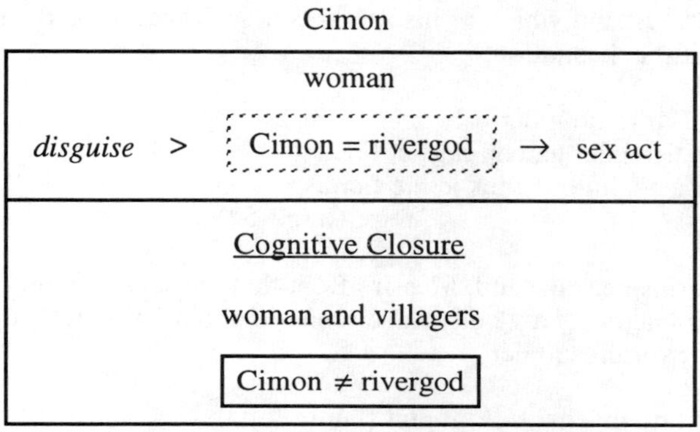

Figure 3.3

In the more complex patterns of deceit accomplished by disguise, the deceivee is manipulated into believing that he (they are all men) is the initiator of the trickery. He is the *trompé se croyant trompeur*. The embedded dupery does not succeed, and, consequently, the intended deceivee's belief is not altered. In such cases—there are four of them—the disguise becomes the very means by which the person using it (albeit unwittingly in "La Gageure des trois commères") is duped by the primary deceivers. A further distinctive feature is that the disguise is intended not to seduce but to conceal ("Le Cocu battu et content," first tale in "La Gageure des trois commères," "Les Lunettes") or to reveal ("Le Mari Confesseur") an illicit sexual relationship. In the first three, the disguises are *travestissements*, a transsexual disguise in which a man dresses as a woman, a type of disguise rare in seventeenth-century theatre (Forestier 472).[6]

"Le Cocu battu et content" is inspired by Boccaccio (VII, 7), but its subject harks back to the medieval topos of the *mari berné* in which the disguise is destined to make its bearer appear ridiculous.[7] A misrepresentation takes place at the outset when a young man, wanting to meet a woman he has just seen, introduces himself to Messire Bon, her husband, as a *fauconnier*[8] looking for employment. Once accepted into the household, he loses little time in making his desires known to the wife of Messire Bon.

The primary deception of the tale operates through disguise and has as its agents the falconer and Messire Bon's wife. Their goal in the deception is to have some time together without the surveillance of the husband, while at the same time confirming the husband's erroneous belief in the continuing fidelity of his servant and his wife. The wife takes the initiative in launching the deceit by asking her husband which of his servants he believes to be the most faithful. He replies without hesitation:

> . . . J'ai toujours cru
> Le fauconnier garçon sage et fidèle;
> Et c'est à lui que plus je me fierais.
> (ll. 52-54)

She then informs her husband, Messire Bon, that his servant is not as faithful as he seems and makes up a wholly false scenario about her reaction to the falconer's amorous overtures to her:

> C'est un méchant: il me tint l'autre fois
> Propos d'amour, dont je fus si surprise,
> Que je pensai tomber tout de mon haut;
> Car qui croirait une telle entreprise?
> (ll. 56-59)

She claims to have pretended to cooperate with the falconer and to have made a rendezvous in the garden under a pear tree (Boccaccio's tree is a pine) that very evening when her husband would be sleeping most soundly, "pendant son premier somme" (l. 71).

As a result of all this false information Messire Bon's perspective on the situation has shifted radically. He now believes—his framing of the situation is doubtful—that his servant is possibly unfaithful to him and in love with his wife. What is particularly ironic about the deception is that the belief space that they invent for the husband corresponds to the real situation that the husband is too blind to see.

Messire Bon's wife persuades him to disguise himself in her clothing and, then, go to meet her lover in the garden:

> Dans le jardin attrapez-le vous-même:
> Vous le pourrez trouver fort aisément;
> Le poirier est à main gauche en entrant.
> Mais il vous faut user de stratagème:
> Prenez ma jupe et contrefaites-vous;
> Vous entendrez son insolence extrême:
> Lors d'un bâton donnez-lui tant de coups,
> Que le galant demeure sur la place.
> (ll. 78-85)

While the husband waits and shivers in the garden, the lover is in *his* bed enjoying forbidden pleasures, "femme jolie, et qui n'est point à soi" (l. 102). This first phase of the plan is easily executed; the husband, waiting in the garden, becomes a *cocu*. The substitution by disguise also goes according to plan from the false perspective of the husband. He, as a pseudoagent in the deception, believes the purpose of the disguise is to unmask the unfaithful servant. He believes there to be a simple pattern of the servant's deception and his own counterdeception. The spell of a pear tree is, however, again at work! Nothing is as it seems.

The second phase of the deception, the actual confrontation in the garden, is developed at greater length than in Boccaccio's tale and with more emphasis on the husband's false belief.[9] The lover pretends to believe—"feignit de croire"—the disguise and feigns loyalty to the husband, flattering him as he berates the supposed wife:

> . . . Quoi donc méchante femme!
> A ton mari tu brassais un tel tour!
> Est-ce le fruit de son parfait amour!
>
> (ll. 112-14)

He had invited her to the garden, so he claims, only to test her fidelity and now pursues her with a stick, beating her as she retreats. Messire Bon, "le cocu battu et content," having approached the situation in a doubtful cognitive frame, is completely duped, and even takes pleasure in the beating. The loyalty of his servant is absolutely assured now in his mind. La Fontaine emphasizes the transformed belief state of the husband:

> Pendant ces mots l'époux pleurait de joie,
> Et, tout ravi, disait entre ses dents:
> Loué soit Dieu, dont la bonté m'envoie
> Femme et valet si chastes, si prudents.
>
> (ll. 130-33)

In Boccaccio's version the husband does not react; he goes back to his wife and recounts briefly what has happened. In La Fontaine's "Le Cocu battu et content," the beating is more severe, but to the delight of Messire Bon, his attempted *trompeur trompé* scheme is unsuccessful. The falconer is not a *trompeur* after all!

> Messieur Bon eût voulu que le zèle
> De son valet n'eût été jusque-là;
> Mais le voyant si sage et si fidèle,
> Le bonhommeau des coups se consola.
> (ll. 138-41)

For the perpetrators of the deceit the disguise has achieved its goal: Messire Bon is firmly anchored in his counterfactual mental space.

The ending of the tale is replete with dramatic and verbal irony as Messire Bon praises the faithful *fauconnier*, hoping that he will "take" a wife in their village, and, finally, encouraging his wife to treat the servant just as she would treat him.

> Quand nous pourrions vivre cent ans encor,
> Ni vous ni moi n'aurions de notre vie
> Un tel valet; c'est sans doute un trésor.
> Dans notre bourg je veux qu'il prenne femme:
> A l'avenir traitez-le ainsi que moi.
> (ll. 144-48)

She, unsurprisingly, promises with great alacrity:

> Pas n'y faudrai, lui repartit la dame;
> Et de ceci je vous donne ma foi.
> (ll. 149-50)

It is the nature of poetry to invest words with connotative power; their repetition becomes highly evocative. In "Le Cocu battu et content" La Fontaine's text plays poetically with the ambiguities and discordant recurrences of *foi* and *fidèle*. The last word of the poem is *foi*, a word that calls up all the semantic resonances associated with this leitmotif.

Boccaccio's conclusion puts less emphasis on the discrepant states of knowledge of the characters. In the *Decameron* the wife, not the husband, makes the final speech about the faithful servant:

> Lodato sia Iddio che egli ha me provata con parole e te con fatti; e credo che egli possa dire che io porti con piu pazienza le parole, che tu i fatti non fai. Ma poi che tanta fede ti porta, si vuole aver caro e fargli onore. Egano disse: Per certo tu di'il vero. (441)

> [Thank God he put me to the test with words and you with deeds. I think he can claim that I submitted more patiently to his speech than you to his action. Still, look at his fidelity: you must show him your appreciation. —You're absolutely right. (450)]

Disguise in this tale creates overlapping contradictory belief spaces. The same situation is viewed simultaneously through different frames of interpretation. The husband believes—in a counterfactual mode—that his servant is fooled by the disguise and that he gives a definite proof of his trustworthiness. The lover and the woman know that the scene in the garden is simulated. Messire Bon's own attempted deception is embedded in the major act of deception in which the wife and lover achieve their intended goal by the same means—disguise of the husband—as Messire Bon uses. Only one belief space, the husband's, is manipulated. He is duped and remains ignorant of his status as a *cocu*. The disguise failed as a means of impersonation since no one was deceived by the false appearance of Messire Bon. Its purpose, though, was a subtler kind of cognitive manipulation—to influence the husband's mental framing of the scene in the garden and thus to confirm his confidence in the falconer. This is a curious case where the person wearing the disguise in fact is being duped by means of the disguise. There is no moment of discovery, the dupery remains concealed, and no punishment ensues.

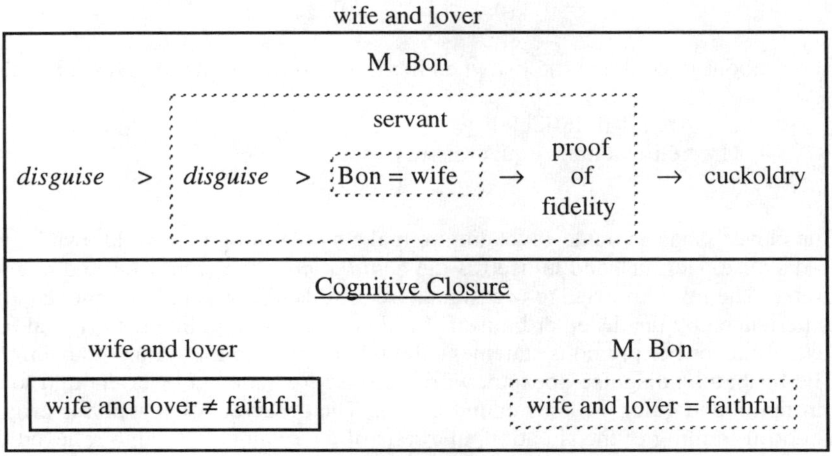

Figure 3.4

In "Le Mari confesseur" Artus, the husband, who has just returned from a long voyage, disguises himself as a priest hearing confessions intent on learning from his wife whether she has been unfaithful during his absence. His mental framing of his wife's activities is operating in the doubtful mode. Greatly abbreviating his source, a tale in the fifteenth-century *Les Cent nouvelles nouvelles* (*Conteurs français* 287-90), La Fontaine suppresses details about the wife's infidelity and chooses to emphasize the husband's quandary in an interior monologue:

> Messire Artus ne prit goût à l'affaire;
> Et ruminant sur ce qu'il devait faire:
> Depuis que j'ai mon village quitté,
> Si j'étais crû, dit-il, en dignité
> De cocuage et de chevalerie:
> C'est moitié trop, sachons la vérité.
> (ll. 14-19)

He strongly suspects that she is unfaithful, and disguises himself as a priest. She presents herself to confess her sins, beginning with the less serious ones, then states:

> Père, dit-elle, en mon lit sont reçus
> Un gentilhomme, un chevalier, un prêtre.
> (ll. 27-28)

She is about to continue the list when her husband interrupts to reveal himself:

> . . . Ah, dit-il, infidèle!
> Un prêtre même! A qui crois-tu parler?
> (ll. 33-34)

She cleverly has a counterrevelation to make by claiming to have known he was in disguise. Her husband himself is the *gentilhomme*, the *chevalier*, and now the *prêtre*. The false knowledge occasioned by the disguise was only momentary and experienced by the deceiver himself! La Fontaine's text, unlike the original *nouvelle*, reduces her previous statement about her lovers to the status of an enigma. The husband's disguise does not work because his wife is clever enough to see through it and quickly plan a counterattack. The deceiver is left with his original doubtful framing of the situation, in a state of uncertainty, although relieved that what he perceives as a momentary counterfactual framing of his belief space has passed (see Figure 3.5 below):

> Béni soit Dieu! dit alors le bon homme:
> Je suis un sot de l'avoir si mal pris.
> (ll. 44-45)

Less satisfactory than most of La Fontaine's tales because of its brevity and its less nuanced mode of presentation, "Le Mari confesseur" is epigrammatic in its lack of detail and in its reliance on wordplay to bring about a resolution to the narrative. In the version of the tale in *Les Cent nouvelles nouvelles*, for example, the husband reacts more volubly:

> Par ma foy, m'amye, dist lors le chevalier, or m'avez vous vaincu et bien monstré que sage et tresbonne vous estes, et que sans cause et a tort et tres mal adverty vous ay chargée et dist du mal assez, dont il me desplaist; et m'en repens, et vous en crye merci, vous promectant de l'amender a vostre dit. (290)

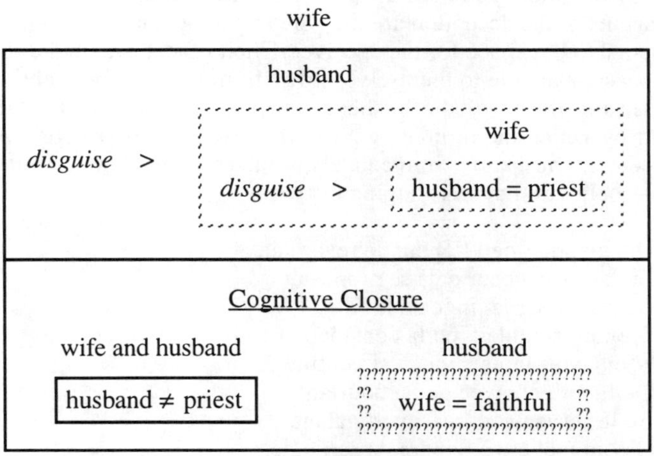

Figure 3.5

The contrast between the prose version with its redundant style and La Fontaine's poetic version with its exaggerated brevity is striking. The medievalist, Roger Dubuis, sees that one could be inclined to judge La Fontaine's version unfavorably as being less dynamic. He points out, however, that in his view what distinguishes La Fontaine's tale is the poetry that adds an elegance and pithiness to the expression: "La Fontaine, maîtreversificateur, parvient à transposer dans un régistre différent la technique très élaborée d'un prosateur qui connaissait bien son métier" (357).

Another instance of embedded deception where the husband unwittingly uses a disguise to set the scene for further trickery of himself occurs in the initial tale of "La Gageure des trois commères." Unlike the situation in "Le Cocu battu et content" and "Le Mari confesseur" the deceivee—again a husband—does not instigate the disguise and is, in fact, completely unaware of it.

This is a tale of sexual ambivalency, possibly invented by La Fontaine,[10] where a woman dupes her husband by disguising her lover as a female servant. Georges Forestier finds that in the theater the numerous instances of men disguised as women often present dubious situations in which a woman is sexually attracted to another female. Even a hint of masculine homosexuality is, however, extremely rare: "Constatons que seuls les personnages féminins peuvent être concernés par des relations équivoques" (Forestier 373).[11] La Fontaine's depiction of this sexual transgression in "La Gageure des trois commères" is accomplished delicately by the interposition of the counterfactual framing of disguise.

The agents of the deception are the wife and her young lover who construct a counterfactual belief space for the deceivee. Their object is to make some private time and space available to themselves away from the closely attentive husband. In this case the libido of the husband becomes the object of attack and the instrument by which the ultimate goal of the deception is realized. As the wife had foreseen, her husband is immediately attracted to the new servant girl, and—ironically—believes that *he* is duping his wife:

> Du coin de l'oeil l'époux le regardait,
> Et dans son coeur déjà se proposait
> De rehausser le linge de la fille.
> Bien lui semblait, en la considérant,
> N'en avoir jamais vu de si gentille.
> On la retient, avec peine pourtant:
> Belle servante, et mari vert galant,
> C'était matière à feindre du scrupule.
> Les premiers jours, le mari dissimule,
> Détourne l'oeil, et ne fait pas semblant
> De regarder sa servante nouvelle.
>
> (ll. 56-66)

The pivotal scene of the deception plan involves a situation that bears some similarity to the situation in "Le Cocu battu et content." Here, too, the husband believes he is duping his wife, whereas, in truth, he is the pawn in a larger game of deception. One night the purported female servant announces that the wife is sick and will sleep alone. The foolish husband tumbles into the trap and takes her to his bed. As she is lying in bed in her nightcap and just about to remove her skirt, the wife bursts into the room:

> Madame vient: qui fut bien empêché
> Ce fut l'époux cette fois pris pour dupe.
> Oh, oh, lui dit la commère en riant,
> Votre ordinaire est donc trop peu friand
> A votre goût; et par saint Jean, beau sire,
> Un peu plus tôt vous me le deviez dire:
> J'aurais chez moi toujours eu des tendrons.
> (ll. 77-83)

The husband is verbally beaten into submission. Through his false framing of the situation he sees the servant as being harshly chastized and punished by being committed to the wife's bed from then on. From the reader's viewpoint the scene bristles with dramatic irony as the wife leads her lover away to bed, stating in front of her husband that the servant will not leave her side night or day:

> Vite, marchons; que du lit où je couche
> Sans marchander on prenne le chemin:
> Vous chercherez vos besognes demain,
> Si ce n'était le scandale et la honte,
> Je vous mettrais dehors en cet état.
> Mais je suis bonne, et ne veux point d'éclat:
> Puis je rendrai de vous un très bon compte
> A l'avenir, et vous jure ma foi
> Que nuit et jour vous serez près de moi.
> (ll. 98-106)

The *chambrière* (again a suggestively named occupation)[12] appropriately sheds one or two tears and prepares with renewed enthusiasm to go about her twofold household duties:

> S'en va jouer un autre personnage;
> Fait au logis deux métiers tour à tour:

> Galant de nuit, chambrière de jour,
> En deux façons elle a soin du ménage.
>
> (ll. 113-16)

The disguise is successful much in the same way as in "Le Cocu battu et content" in that it permits the establishment of a new situation in the household by inducing in the deceivee a false frame of interpretation. Like Messire Bon, this husband is blissfully unaware of the presence of his wife's lover. This state of opacity at the conclusion is a final stage in a deception where there is no moment of recognition:

> Le pauvre époux se trouve tout heureux
> Qu'à si bon compte il en ait été quitte.
> Lui couché seul, notre couple amoureux
> D'un temps si doux à son aise profite.
>
> (ll. 117-20)

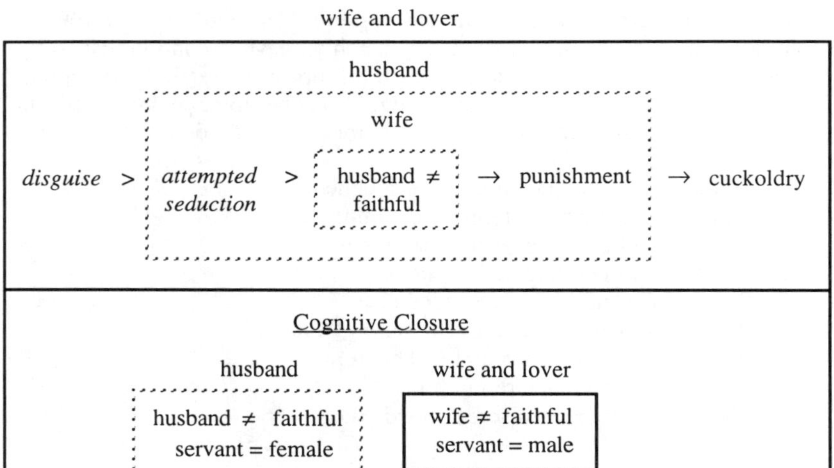

Figure 3.6

In "Les Lunettes" the disguise itself backfires, so to speak, on its author. The tale is inspired by Bonaventure Des Périers's tale, "Du jeune garcon qui se nomma Thoinette pour estre receu à une religion de nonnains; et comment il fit sauter les lunettes de l'Abbesse qui le visitoit" (*Conteurs français* 493-95). In the first part of the tale a young man disguises himself as a nun, soeur Colette, in order to live in the convent with Soeur Agnès. Here, once more, disguise operates not to seduce, but to hide the existence of the lover. The secrecy of the disguise is, though, severely threatened when Soeur Agnès produces a baby (a detail invented by La Fontaine):

> Voilà scandale et bruit dans l'abbaye.
> D'où cet enfant est-il plu? comme a-t-on,
> Disaient les soeurs en riant, je vous prie,
> Trouvé céans ce petit champignon?
> Si ne s'est-il après tout fait lui-même.
> (ll. 33-37)

The prioress frames the incident in the doubtful mode, her highly suspicious interpretation translating itself into a flurry of question marks in the text as she ponders on the existence of a father and wants to *know* the truth:

> Comment est-il entré? comment sorti?
> Les murs sont hauts, antique la tourière,
> Double la grille, et le tour très petit.
> Serait-ce point quelque garçon en fille?
> Dit la prieure, et parmi nos brebis
> N'aurions-nous point, sous de trompeurs habits,
> Un jeune loup? Sus qu'on se déshabille:
> Je veux savoir la vérité du cas.
> (ll. 42-49)

The direct discourse—completely lacking in the original—coupled with the sprightliness of the verse form contrasts with the terseness of La Fontaine's source: "L'abbesse, esbahie de cette nouvelle, en voulut sçavoir la vérité au premier jour, et pour ce faire, fit venir toutes les religieuses en chapitre" (494). La Fontaine's text is also more dramatic at the outset than Bonaventure Des Périers's, partly because the events are slightly changed; in the original the nuns were only spreading rumors that there might be a man in the convent, but in La Fontaine's tale, the baby's arrival precipitates, without any forewarning, an investigation.

The episode that is the focal point of the tale and inspires its title occurs when the prioress orders all the nuns to disrobe. Even during this ultimate scrutiny by the prioress [his "buff" is called, one might say!] the young man somehow ingeniously disguises his telltale features—until, seeing himself surrounded by naked women, he has an erection (La Fontaine is more subtle in describing this event) and knocks off the mother superior's glasses in an uncontrollable movement underlined poetically by the enjambment of lines 113-14:

> Touffes de lis, proportion du corps,
> Secrets appas, embonpoint, et peau fine,
> Fermes tétons, et semblables ressorts,
> Eurent bientôt fait jouer la machine.
> Elle échappa, rompit le fil d'un coup,
> Comme un coursier qui romprait son licou,
> Et sauta droit au nez de la prieure,
> Faisant voler lunettes tout à l'heure
> Jusqu'au plancher. . . . .
>
> (ll. 106-14)

The initial disguise, adopted so as to have access to a woman by deceiving the other nuns, was successful for some time. The final attempt at disguise is, however, short-lived and brings the tale to its cognitive climax. The embedded pattern of false belief frames once more appears; the young man falsely believes that his disguised penis will deceive the prioress. It is, nevertheless, the very attempt at disguise that results in his discovery and the end of his sexual activity.

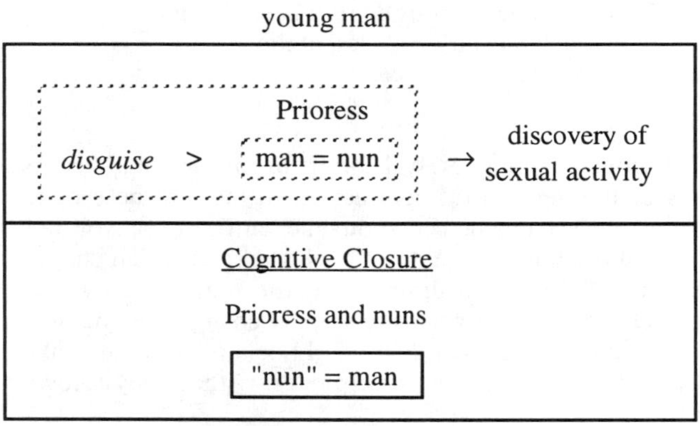

Figure 3.7

As in "Le Muletier" an attempt to mete out punishment is made, but it is foiled by the cleverness of the deceiver. In this instance he persuades a naive young miller to take his place, tied up to a tree, awaiting a beating from the nuns.

***

Disguise, a nonverbal strategy of impersonation, is a false communication fundamentally reducible to a latent structure of substitution. The transgression of the act of communication consists in a deceptive exchange that sets up false equivalencies between the misappropriated semiotic signifiers and the signified personal identity. In La Fontaine's *Contes* subversions of the exchange process generate patterns of deceit within the cognitive space of the diegetic universe. If, as Philippe de Lajarte has pointed out, exchange is fundamental to the structure of all *contes facétieux*, the tales of La Fontaine with their playful emphasis on the multiple patterns of subversive deceit belong to a narrative universe governed by the same "universal law":

> L'univers de l'echange tel que le représente le conte facétieux se subdivise en une *pluralité* d'instances (économique, sociale, religieuse, sexuelle, etc.) dont chacune est régie par un système d'échange spécifique (qui constitue la *loi* de cette instance); elles se divisent en instances *légales* ou dominantes (expressions de diverses idéologies officielles) et en instances *subversives* ou dominées (ex. le sexe, le "*bas matériel*", l'Enfer, etc.) enfin—et bien qu'ayant une tendance à se constituer chacune en un système indépendant et absolu—elles entretiennent entre elles des rapports d'échange, rapports régis par une *métaloi* de l'échange (de l'équivalence) qui constitue, en définitive, l'unique loi universelle. (330)

The specificity of La Fontaine's *Contes* is in the generation, as Lajarte suggests, of its own independent system of substitution and exchange, a fictive world with its distinctive patterns of deceit.

NOTES

1. Forestier finds *l'approche*—gaining access to another person—as the purpose of disguise in 35% of his corpus; two typical examples are Basile in Odet

de Turnèbe's *Les Contents* (1584) and Adraste in Molière's *Le Sicilien* (1668). For another analysis of disguise in a seventeenth-century text, see Gregorio, *The Pastoral Masquerade: Disguise and Identity in "L'Astrée."*

2. The sentence I quote from Boccaccio is translated by Guido Waldman, with a certain delicacy, as follows: "He took her passionately in his arms and pretended to be in a bad temper." This translation of the *Decameron* will be used throughout the text.

3. The word *déduit* was a common euphemism, according to Pierre Guiraud's *Dictionnaire érotique* where it is listed as "*déduit*-coït." Collinet's note to the word in the Pléiade edition of the *Contes* is more subtle, restricting itself to quoting Richelet's definition: "Passe-temps, plaisir. Il est du style le plus simple."

4. Collinet (1494) points out that La Fontaine's attention may have been drawn to this letter by a passage in Fontenelle's *Dialogues des morts* (1683). A French translation of Eschine's text is found in the Régnier edition of La Fontaine, vol. 6, 361-62.

5. The seduction is brief and more brutal in the source: "'Scamandre, dit-il, reçoit le présent de Callirrhoé; il veut la combler de biens.' En disant ces mots, il enlève la fille et se cache."

6. Forestier finds only 3 *travestissements* of a man as a woman after 1640, 16 in all over 130 years (472).

7. Some remnants of the tradition persist in Molière's *Le Médecin malgré lui* and in *Monsieur de Pourceaugnac*, in both of which the disguise is imposed by another person. For medieval examples, see Montaiglon and Raynaud, *Recueil général et complet des fabliaux*: "La Bourgeoise d'Orléans" and "La Dame qui fit battre son mari" (Fab. I, 8 and IV, 100); and both medieval and seventeenth-century analogues in Goran Börnas, "Le Cocu battu et content."

8. Could this be sexual wordplay? In Boccaccio he is simply a servant in the household. According to Guiraud's *Dictionnaire érotique*: "*faucon*—sexe de la femme." Muscatine, in his study of *The Old French Fabliaux*, finds a sexual pun in the medieval fabliau *Guillaume au faucon*: "The elegant courtliness of *Guillaume au faucon* is nearly destroyed near the end by a pun on *faucon* (falcon / false cunt); the author pointedly insists on it as 'two words in one' (*deux moz a un*)" (115).

9. In Boccaccio the whole garden scene is very brief:

Anichino levatosi e nel giardino andatosene con un pezzo di saligastro in mano, come fu presso al pino, ed Egano il vide venire, e cosi levatosi come con grandissima festa riceverlo volesse, gli si faceva incontro; al qualé Anichin disse: -Ahi! malvagia femina, adunque ci se'venuta, ed hai creduto che io volessi o voglia al mio signore far questo fallo? Tu sii la mal venuta per le mille volte!- Ed alzato il bastone, senza dir parola comincio a fuggire, ed Anchino appresso, sempre dicendo: -Via, che Iddio vi metta in malanno, rea femina, ché io il diro domattina ad Egano per certo. (440-41)

[So Anichino got up and went into the garden clutching a length of white willow. As he approached the pine, Egano saw him coming and went to meet him, showing every sign of welcome—but Anichino said: "Why you strumpet! So you've come, have you? Did you believe I really would serve my master such a turn? You make me sick to look at you!" He raised his stick and laid into him but these words and the sight of the cudgel sent Egano fleeing without a word, and Anichino dashed after him, crying: "Be off with you, trollop! Believe me, I'll tell Egano in the morning." (449)]

10. No specific source has been found for the tale—see Collinet's edition of the *Contes*, 1389.

11. Only two occurrences of men falling in love with men disguised as women are cited by Forestier; they are two older kings, one in *Eurimédon ou l'illustre pirate* (1637) of Desfontaines and the other, in *La Cour bergère* (1640) of Mareschal.

12. See Guiraud, *Dictionnaire érotique*: "*chambre*—sexe de la femme" and "*chambrer*— garder en chambre pour faire l'amour."

Chapter Four

Deceitful Topography: Patterns of Displacement, Misplacement, and Replacement

While disguise is a most effective means of impersonation, the process of misidentification can occur in other ways. In disguise the visual appearance is altered; an equally effective strategy of deceit, however, is the manipulation of the actual physical space occupied by a person or an object. In many situations, the place acts as a strong identifying marker of personal identity. In the *Contes* impersonation without disguise operates when someone usurps a physical space where another is expected to be—as, for example, when a wife, under cover of darkness, substitutes herself for the mistress at a rendezvous with her own husband. Such impersonations need not involve disguise. The deception is created by a physical substitution or exchange—someone takes over the place that is known to be reserved for another person. In "Joconde" the two men who are in bed with the innkeeper's daughter are fooled by a substitution by displacement. All night long they each believe that the other is occupied making love to the young woman, whereas, in reality her lover has displaced them both. This pattern of deceit is the major mental space-builder and the chief operator of deceit in a number of La Fontaine's *contes*. Misplacement is an unconscious act that has disruptive consequences in the cognitive domain; in two tales misplacement causes serious misunderstandings. On the cognitive level, displacement and misplacement function like disguise in that the deceit, whether intentional or not, creates a counterfactual space. As for replacement, this is a term I reserve for a few tales in which there is a measure of self-delusion bringing about a dubious state of psychological myopia.

TRADING PLACES: DISPLACEMENT STRATEGIES

In "Richard Minutolo," a tale whose source is Boccaccio's *Decameron* (III, 6), a woman is duped into believing she is displacing her husband's mistress at an assignation, while, unknown to her, her action is merely assisting the main displacement and her own seduction. Richard Minutolo's initial lie to her about the supposed illicit rendezvous between her husband and his mistress remains present in the background of the tale, facilitating the inception and the implementation of the major deception. Catelle falsely believes that this simple substitution of herself for the mistress in "la chambre noire" (l. 71) will result in the confusion and embarrassment of her husband. Minutolo, who suggested her duplicity, constructs his own *trompeur trompé* scheme and is there waiting for her in the dark.

> Là ne trouva ce qu'elle allait chercher:
> Point de mari, point de dame Simone;
> Mais au lieu d'eux Minutol en personne,
> Qui sans parler se mit à l'embrasser.
> (ll. 108-11)

He takes malicious pleasure in making love to her while she all the time believes he is her husband:

> Il en profite, et se garde de rire;
> Mais toutefois ce n'est pas sans effort.
> De figurer le plaisir qu'a le sire,
> Il me faudrait un esprit bien plus fort.
> Premièrement il jouit de sa belle;
> En second lieu, il trompe une cruelle.
> (ll. 118-23)

The ironic phase of Catelle's counterfactual framing of the situation is prolonged as she "reveals"—so she believes—her identity to her husband and indulges in a lengthy and violent diatribe (ll. 126-56) against his infidelity:

> C'est trop souffrir, traître, ce lui dit-elle,
> Je ne suis pas celle que tu prétends.
> Laisse-moi là; sinon à belles dents
> Je te déchire et te saute à la vue.
> (ll. 126-29)

The recognition scene is highly dramatic as Catelle, no longer able to contain her anger, even threatens to make love with Minutolo, and then flings open a window, only to discover Minutolo, not her husband. Richard Minutolo's excuses may seem feeble from a twentieth-century female perspective, however, as Jean-Pierre Collinet reassures us in a lengthy note to this passage (1355), the motif that "Tout est permis en matière d'amour" (l. 169) was generally accepted in the seventeenth century and frequently expressed by other authors as well.

Continuing to indulge in his perverted sense of pleasure, Minutolo becomes even more sexually aroused when she begins to cry:

> Elle se mit à pleurer tendrement.
> En cet état elle parut si belle,
> Que Minutol de nouveau s'enflammant,
> Lui prit la main . . . .
> (ll. 180-83)

He in fact manages to convince her that there is no further harm to be done by continuing their actions. After what might be termed the rape of Catelle, her true seduction takes place with her gradual consent, first to listening, then to a kiss, and finally "[elle] vient à son point" (l. 214), an expression more directly explained by Pierre Guiraud than by Collinet's anodyne quotation from the *Dictionnaire de l'Académie française*: "le faire condescendre à ce qu'on souhaite."[1] La Fontaine mutes the shocking effect of Minutolo's sadistic attitude by commenting that after Catelle becomes a willing partner, Richard's own pleasure is augmented: "Heureux cent fois plus qu'il n'avait été!" (l. 215). And, what is more, the narrator further strengthens the process of attenuation by adding, in a personal aside to the reader:

> Car, quand l'amour d'un et d'autre côté
> Veut s'entremettre, et prend part à l'affaire,
> Tout va bien mieux, comme m'ont assuré
> Ceux que l'on tient savants en ce mystère.
> (ll. 216-19)

The large frame (see Figure 4.1 below) belongs to Richard Minutolo, the chief manipulator, who is simultaneously aware of Catelle's counterfactual belief space—which he has set in motion—and what he knows is the factual situation. The box with the broken line representing Catelle's counterfactual belief space contains within it another frame representing what she falsely believes to be her husband's counterfactual space. This is an embedded deceit pattern.

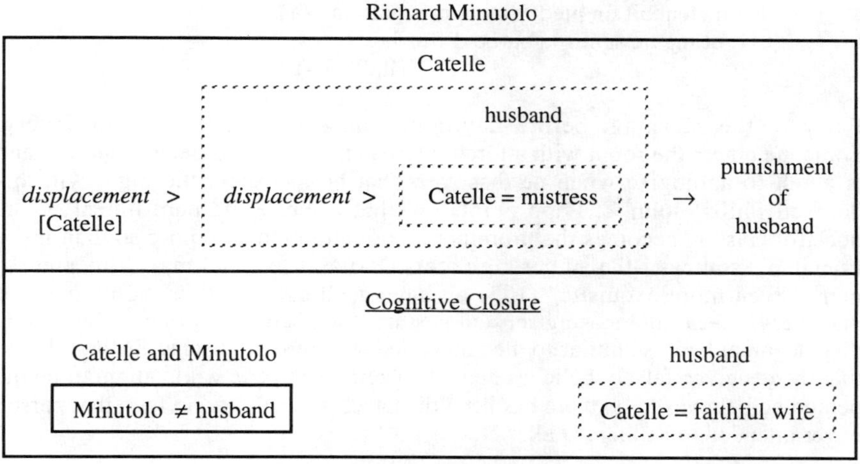

Figure 4.1

Catelle falsely believes that she is displacing her husband's mistress and that the displacement is deceiving her husband; at the same time she falsely believes that what is really going on is a sexual encounter with her own husband. She believes that her husband is present and that he mistakes her for his mistress. Richard Minutolo knows that the true displacement operative in the trickery is the second one, his usurpation of the husband's place. The pleasure he takes in the sexual act is enhanced by his ironic juxtaposition of cognitive frames, his own and Catelle's.

A displacement creating a cognitive pattern similar to the one in "Richard Minutolo" occurs in "Le Gascon puni" where Philis, intent on teaching her overzealous suitor, Dorilas, a lesson, succeeds in manipulating him into doing a special favor for her. She constructs a totally false belief space for him. He believes that he is displacing Cloris, the woman next door, in her bed that night because she needs to see her lover, Damon. Dorilas, in bed with someone whom he believes to be Cloris's husband, spends an excruciating night in continual fear of discovery:

> C'est qu'il craignait qu'enfin un caprice amoureux
> Ne prît à ce mari: tels cas sont dangereux,
> Lorsque l'un des conjoints se sent privé du somme.
> Toujours nouveaux sujets alarmaient le pauvre homme:

L'on étendait un pied; l'on approchait un bras:
Il crut même sentir la barbe d'Eurilas.
(ll. 79-84)

Early in the morning, before daylight, comes the moment of discovery. Someone enters the room with a torch. Dorilas thinks he has been found out and is about to apologize when he discovers that he has spent the night with his beloved Philis! John C. Lapp points out that because the punishment of the boastful Gascon becomes the prominent motivation—much more so than in La Fontaine's sources—the author highlights Dorilas's fear and makes his punishment "even more exquisite" (133) by having Philis, just after the moment of discovery, reveal in a teasing sensual gesture, "son sein à demi nu." The double displacement here is similar to the embedded structure of "Richard Minutolo" in that the deceivee falsely believes he is displacing someone while all the time the person he believes to be there has been displaced as well—or, in fact, that person never intended to be there at all.

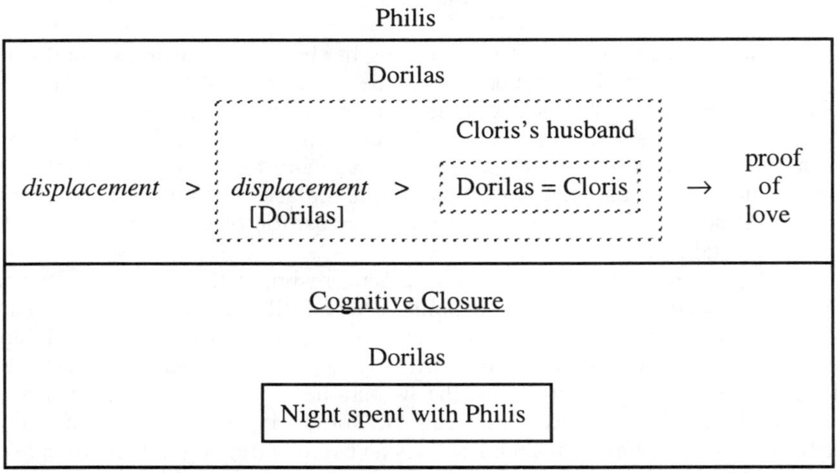

Figure 4.2

The deceiver, as in the previous tale, delights in the privileged knowledge of the larger frame—viewing Dorilas's understanding of the situation as opposed to her own.

"Les Quiproquos," published posthumously in 1696, provides yet another instance of the "bons tours des maris et des femmes" (l. 499) that were presaged in "Joconde," the very first tale of the *Contes*. The tale's subject, adapted primarily from Marguerite de Navarre's *Heptaméron* (VIII), and its prefatory comments on the game of substitution make it a particularly fitting postscript to the *Contes*. In the first lines the narrator views those incidents in which one person is substituted for another as instances of the games Fortune plays on the human race:

> Dame Fortune aime souvent à rire,
> Et nous jouant un tour de son métier,
> Au lieu des biens où notre coeur aspire,
> D'un quiproquo se plaît à nous payer.
> Ce sont ses jeux, j'en parle à juste cause.
>
> (ll. 1-5)

The narrator presents the tale he is about to recount as a typical example of an exchange subverted by a displacement:

> Disons un troc, où réciproquement
> Pour la soubrette on employa la femme.
>
> (ll. 21-22)

Clidamant is married to a beautiful and virtuous woman, but out of curiosity he proposes a sexual encounter to his wife's servant, Alix. Finally, after the sum of "cent beaux écus" has been agreed upon as a fair exchange, Alix agrees to meet Clidamant in "la cave." The two women, however—the wife and Alix—join forces to dupe the husband by making a substitution of the wife for the servant. Unknown to them, Clidamant makes a deal with another man to share the favors of Alix, counting on the darkness to conceal the displacement: "dans l'obscurité / Alix serait fort aisément trompée" (ll. 114-15). The two men go to the rendezvous in the dark expecting that Alix will not even know that there are two of them. The disorienting effect of the dark room is expressed by the suddenly disjointed rhythm of the text:

> On ouvre, on entre, et sans retardement,
> Sans lui donner le temps de reconnaître
> Ceci, cela, l'erreur, le changement,
> La différence enfin qui pouvait être
> Entre l'époux et son associé,

Avant qu'il pût aucun change paraître,
Au dieu d'amour il fut sacrifié.
(ll. 146-52)

Clidamant only discovers the substitution of his wife for Alix when it is too late. Both husband and wife were naively functioning within false belief frames—embedded structures of deceit—the husband believing he was making love to Alix, the wife believing she was making love solely with her husband, although she did wonder about his unusual enthusiasm:

A peine fut cette scène achevée.
Que l'autre acteur, par la prompte arrivée
Jeta la dame en quelque étonnement;
Car comme époux, comme Clidamant même,
Il ne montrait toujours si fréquemment
De cette ardeur l'emportement extrême.
(ll. 157-62)

The moment of recognition is alluded to summarily as occurring when they all emerge "du noir séjour." La Fontaine omits the whole recognition scene in Marguerite de Navarre's tale where the husband discovers his wife's ring on his companion's finger and then confronts her with the evidence. In the concluding lines the narrator-author seems more preoccupied with the moral and epistemic problems of whether Clidamant is a *cocu* or not and the extent of Alix's guilt. His wife did make love to another man, but she thought he was her husband. From a sociological viewpoint, such tales about sexual escapades in which a woman unknowingly enjoys forbidden sex seem to arise from yearnings for guiltless sex in an age where according to the prevalent moral code a sin was only grave if there was sufficient knowledge. The cognitive framing of transgressions was crucial to the ethical debate (see Figure 4.3 below). Alix and Clidamant's wife falsely believe that only one displacement is operative. Juxtaposed to their false belief is the counterfactual space of the two men who believe that they have deceived Alix. What distinguishes this cognitive pattern from the one in the two previous tales, "Richard Minutolo" and "Le Gascon puni," is not only the juxtaposition pattern, but the fact that here there is no larger all-encompassing frame delegated to an observer within the diegetic world. In the diagram below for "Les Quiproquos" no unbroken line indicates the presence of a knowledgeable observer. Until the recognition scene only the narrator and the reader appreciate the irony of the naive juxtaposed cognitive spaces.

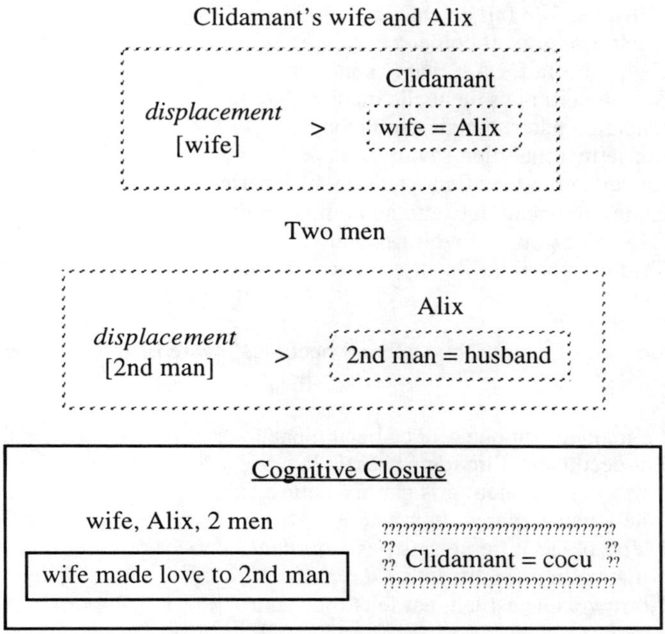

Figure 4.3

The next three tales present displacement as a defensive strategy. In 'La Servante justifiée," a highly original tale that owes only a few details to Marguerite de Navarre's *Heptaméron* (XLV), two scenes are superimposed. In the first a neighbor sees the husband next door making love to a female servant in the garden. The steps of the seduction ritual—the portrayal of the servant plucking a bouquet of flowers, the gesture of the husband's hand on her breast, the flowers thrown at the would-be lover, and her reluctant acceptance of his "favors"—are evoked by a playfully poetic text rich in serpentine sibilants, alliterations of *f* accompanied by wordplay with variations on the words *fait* and *fête*, and metonymic flowers:

> Elle faisait un bouquet pour madame:
> C'était sa fête. Voyant donc de la femme
> Le bouquet fait, il commence à louer
> L'assortiment; tâche à s'insinuer:

S'insinuer, en fait de chambrière,
C'est proprement couler sa main au sein:
Ce qui fut fait. La servante soudain
Se défendit: mais de quelle manière?
Sans rien gâter: c'était une façon
Sur le marché; bien savait sa leçon.
La belle prend les fleurs qu'elle avait mises
En un monceau, les jette au compagnon.
Il la baisa pour en avoir raison:
Tant et si bien qu'ils en vinrent aux prises.
(ll. 25-38)

In an ironic turn of events the husband becomes aware of the neighbor's cognitive gaze:

Le malheur fut que tout ce beau ménage
Fut découvert d'un logis près de là.
Nos gens n'avaient pris garde à cette affaire.
Une voisine aperçut le mystère.
L'époux la vit, je ne sais pas comment.
Nous voilà pris, dit-il à sa servante.
Notre voisine est languarde et méchante.
(ll. 41-47)

In the cognitive dimension of the text, the husband must control the circulation of knowledge from the neighbor to his wife. He must construct a false frame of interpretation through which his wife will interpret the information erroneously when it is communicated to her by the neighbor. By reenacting, albeit with less enthusiasm, the same game of seduction with his wife, he deceives her into believing that there was only one scene in the garden and that she was the woman involved. The dramatic nature of the confrontation with the neighbor is spotlighted as the text momentarily transforms itself into the dramatic mode. The first speech—as the text moves into a narrative replay of the scene—reactivates the *fleurs* metonymy and the *fête-fait* motif:

LA VOISINE
Voire! écoutez le reste de la fête:
Vous ne savez où je veux en venir.
Les bonnes gens se sont pris à cueillir
Certaines fleurs que baisers on appelle.

LA FEMME
C'est encor moi que vous preniez pour elle.
(ll. 81-85)

The wife's fivefold testimony, "C'était moi," is convincing, at least to herself. The displacement in this tale is more a displacement of one event by another than a displacement of one person by another. It also differs from other tales in that the deception is meant to alter the deceivee's interpretation of a *past* event. This is like the magician's sleight of hand. The illusion is very effective, although at the end the neighbor is not convinced and the ironic pun on *servie* makes for a complex cognitive closure:

LA VOISINE
Ah! c'est assez. Je ne m'informe plus:
J'ai pourtant l'oeil assez bon, ce me semble:
J'aurais juré que je les avais vus
En ce lieu-là se divertir ensemble.
Mais excusez; et ne la chassez pas.

LA FEMME
Pourquoi chasser? j'en suis très bien servie.

LA VOISINE
Tant pis pour vous: c'est justement le cas.
Vous en tenez, ma commère m'amie.
(ll. 100-7)

In the double tale, "Le Roi Candaule et le Maître en droit," the second narrative (ll. 123-350) presents an attempted displacement being adroitly foiled by a counterdisplacement. A young French law student in Rome is initiated by his teacher into an infallible method of arranging a tryst with a married woman—all Roman women understand, he says, that a young man offering them holy water from the font at the entrance to a church is seeking an amorous adventure. The advice works remarkably well, but a problem arises when the teacher suddenly realizes that the unfaithful woman is his own wife! The knowledge frame of the *maître en droit* suddenly expands with this new knowledge:

. . . Le pis fut que mon homme,
En s'informant de tout, et des si et des cas,

Et comme elle était faite, et quels secrets appas,
Vit que c'était sa femme en somme.
(ll. 232-35)

The narrator realizes that even the cognitive belief space of the reader needs reinforcement:

A ce discours, ami lecteur,
Vous ne croiriez jamais, sans avoir quelque honte
Que l'héroïne de ce conte
Fût propre femme du docteur.
Elle l'était pourtant . . . .
(ll. 228-32)

The projected substitution of the teacher for the student consists in his displacing the young man at the next rendezvous so as to teach his wife a lesson. The darkness, however, does not prevent his discovery, and the perceptive *entremetteuse* recognizes him and circumvents the substitution. She engineers a counterattack and humiliates the law instructor publicly. His false belief that the substitution is working leads him to cooperate fully: she insists that he disrobe, "Madame attend au lit" (l. 317), and when she leads him naked through a darkened labyrinth of corridors and finally through a door—not into the bedroom, but into a classroom where the law students are assembled—the place itself is displaced!

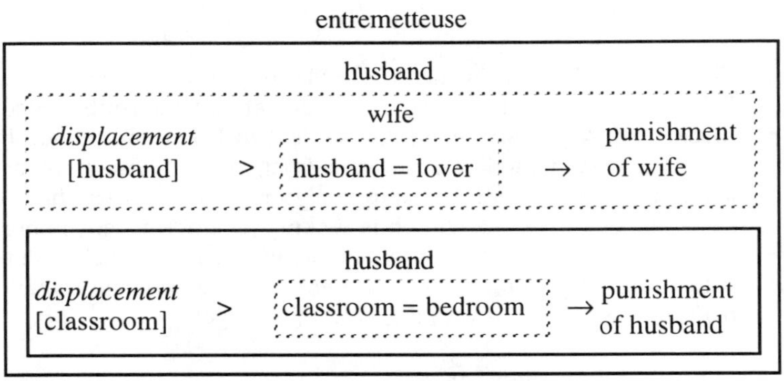

Figure 4.4

In "Le Remède" the displacement does not occur in the dark, but operates by hiding the identifying parts of the body. A young girl sleeping in her room with her fiancé is abruptly awakened by her governess who wants to give her an enema. The lover hides under the covers, as does the confused girl:

> La fille alors ne fut pas assez fine;
> Elle n'avait qu'à tenir bonne mine,
> Et faire entrer l'amant au fond des draps,
> Chose facile autant que naturelle:
> L'émotion lui tourna la cervelle;
> Elle se cache elle-même, et tout bas
> Dit en deux mots quel est son embarras.
>
> (ll. 81-87)

Then the lover solves the problem by presenting the required part of his anatomy to the nurse—*he* gets the enema. This is a simple pattern of displacement, but curious in that it is not the usual kind where one person displaces another who is absent. The girl is present and only one part of her is displaced—could this be termed a narrative synecdoche? The substitution of the part for the whole?

MISPLACED OBJECTS

In the *Contes* an act of deception effected by a schema of misplacement differs from deception by displacement or disguise chiefly in that there is no conscious deceiver. Philosophers, psychologists, and biologists debate about whether deception can be unintentional. It would appear that in the animal world various strategies, such as camouflage and mimicry, are neither intentional nor conscious; the chameleon conceals itself by changing color, and the firefly imitates the light signals of other species. In the human world, too, whether intentionality is essential to the act of deception is a matter of debate.[2] I contend that in La Fontaine's *Contes* there are instances of unintentional deceit through "misplacement," that is, an unconscious act of putting an object in a place where it should not be, which results in an unintended sexual transgression or the inadvertent revelation of a sexual transgression.[3]

In "Le Berceau" the deceptive action is triggered by two deceivees who unintentionally engender their own counterfactual belief spaces. The tale has its source in Boccaccio (*Decameron* IX, 6), but La Fontaine revitalizes it in his own manner: "la mise en vers, par ses exigences et ses ressources, a permis dans le détail un renouvellement complet" (Collinet 1379).[4]

Pinucio invents a stratagem to overcome formidable parental obstacles to being with his beloved Colette. He and a friend arrive at her father's inn so late one night that they are obliged to sleep in the same room as Colette, her parents, and their baby who is sleeping in a cradle next to the parents' bed. Pinucio's plan involves a simple scheme of sleeping with the daughter without her parents' knowledge. A complex game of musical beds takes place, however, in the middle of the night. The friend has to get up, but in order to open the door he must move the cradle close to his own bed. When he returns to bed, he neglects to restore the cradle to its original position. By this time his sleeping companion, Pinucio, has deserted their bed for his fiancée's. The mother then hears a noise and gets up to investigate. When she returns, she mistakenly gets into the friend's bed, misled by the position of the cradle. Somewhat later, Pinucio leaves Colette's bed and mistakenly climbs into bed with her father. The misplacement of the cradle has led to the creation of counterfactual belief spaces, in varying degrees, for all of the characters: the wife who thinks she is sleeping with her husband, Pinucio who thinks he is in bed with his friend, the daughter who thinks Pinucio is back in bed with his friend, the friend who thinks Pinucio is still with the daughter, the husband who, even in the end, never discovers the truth about his wife.

A dramatic disclosure takes place when Pinucio, believing he is addressing his friend, whispers to Colette's father:

> Ma foi! Colette est un morceau de roi.
> Si tu savais ce que vaut cette fille!
> J'en ai bien vu, mais de telle, entre nous,
> Il n'en est point. C'est bien le cuir plus doux,
> Le corps mieux fait, la taille plus gentille;
> Et des tetons! je ne te dis pas tout.
> (ll. 139-44)

La Fontaine's depiction of Colette's beauty is more sensual than Boccaccio's in which Pinuccio only praises her as a "dolce cosa." The moment of discovery is upon them; as the husband reacts, the counterfactual belief frames of the others disappear:

> Quoi! ne tient-il qu'à honnir des familles?
> Pour vos ébats nous nourrirons nos filles,
> J'en suis d'avis. Sortez de ma maison:

Je jure Dieu que j'en aurai raison.
Et toi, coquine, il faut que je te tue.
(ll. 162-66)

The wife, who has actually enjoyed her stay in the friend's bed—all the while somewhat surprised within her false frame of belief at the unusual vigor of her supposed husband—must now save her reputation. She upholds the friend's story that Pinucio was sleepwalking and dreaming, and then adds that he must have been in his own bed because she slept all night long with her daughter. At the end only the husband is left in a false belief space.

The "si tu savais" (l. 140) hovers over the narrative creating ironic cognitive closure. Knowledge about the transgressions of Pinucio, the daughter, the wife, and the companion is not shared with the husband: "Car chacun d'eux en avait sa raison. / Tout fut secret" (ll. 198-99). The unwitting misplacement of the cradle brought about two further misplacements—of the wife and the daughter's lover—by which they unintentionally occupy alien places and engender their own false belief spaces. Until the moment of revelation, the woman believes she is in her husband's bed; the lover believes he is in his friend's bed. The pattern of deceit in the following diagram illustrates naive juxtaposed frames; there is no deceiver and no knowledgeable observer within the storyworld.

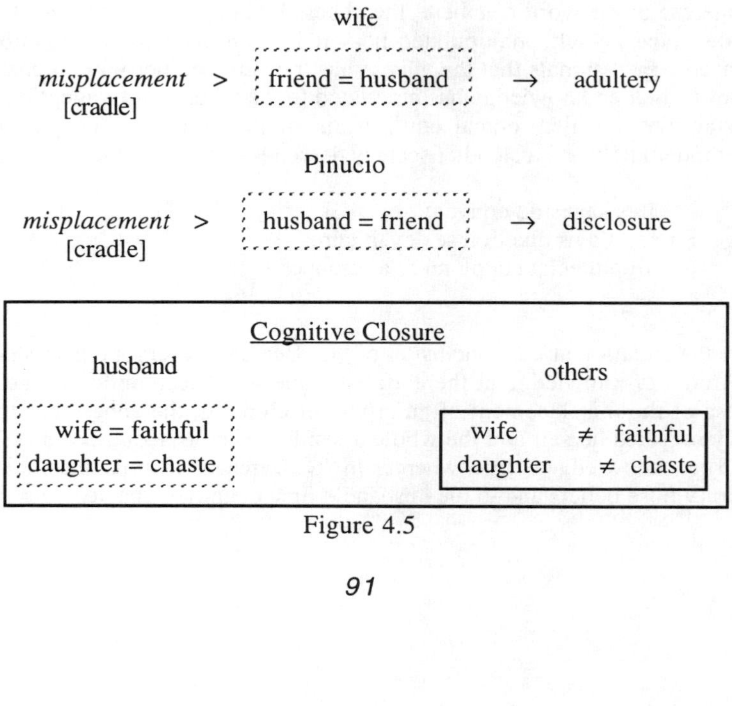

Figure 4.5

Inspired by Boccaccio (*Decameron* IX, 2), "Le Psautier" presents another unintentional deception that is precipitated by the misplacement of an object. Unlike "Le Berceau," where the misplacement caused secretive and unintentional sexual transgressions, the substitution in this tale reveals a secret, thereby allowing an intentional sexual transgression to go unpunished. Sister Isabelle has a man in her convent cell. When the other nuns denounce her, the abbess dresses hurriedly and mistakenly puts a man's pant on her head instead of a veil (her *psautier*). Much like an unconscious Freudian slip, this involuntary act reveals a secret she has wanted to hide; she, herself, has been entertaining Messire Jean, the curé, in her own bed! When Sister Isabelle appears in chapter to be judged by the abbess and receive her penance before the assembled nuns, she notices the strange headdress and remarks: "Votre psautier a ne sais quoi qui pend; / Raccommodez-le" (ll. 103-4). The abbess is humiliated by her unintentional substitution:

> La voix manquant à notre sermonneuse,
> Qui de son troc bien fâchée et honteuse
> N'eut pas le mot à dire en ce moment.
> (ll. 125-27)

Note the use of the word *troc* here; the abbess by her substitution has, if not exactly deceived herself, manipulated herself into an undesirable situation. The misplaced object signals that the abbess has transgressed her vow of chastity. A piece of forbidden knowledge has circulated to the whole entourage, threatening to destabilize the hierarchical equilibrium of the convent. She puts off the chapter indefinitely and a cloud of secrecy descends over the convent:

> . . . Les sages du couvent
> Furent d'avis que l'on se devait taire;
> Car trop d'éclat eût pu nuire au troupeau.
> (ll. 136-38)

Unlike the situation at the conclusion of "Le Berceau" everyone here shares the same frame of knowledge at the end; each one is an accomplice in the secret. Because of the misplacement of an article of clothing, the abbess has unintentionally revealed herself and the whole assembly of nuns acquires a new knowledge. Their knowledge is true, whereas in "Le Berceau" the misplacement led to temporary false beliefs and to the husband's final cognitive opacity.

REPLACEMENT AND SELF-DELUSION

The tales in which deception unfolds according to a schema of replacement contain a prominent component of self-deception. The characters know they are making a substitution, and in this they differ from the unconscious agents of substitution in "Le Berceau" and "Le Psautier." They do not know, however, that the substitution they are making is delusionary. The unintentional character of their acts raises the epistemic problem of self-deception.

Philosophers are intrigued by the apparent paradox of self-deception.[5] Is self-deception, in fact, possible? How can one lie to oneself? Is self-deception simply a species of wishful thinking? Can it be explained by a homuncular theory according to which the human psyche is composed of subsystems of emotions and irrational beliefs, any one of which can interfere with the principal system of rational belief? What role does pseudorationalization play in what we call self-deception?

The tale recounted in the first part of "Le Roi Candaule et le Maître en droit" (ll. 1-116) can be traced back to Herodotus and has frequently been retold.[6] It exhibits an embedded deceit structure in which the king unwittingly brings about his own undoing. Mesmerized by the sight of the queen, he has an irrepressible urge to exhibit her to Gygès:

> Vous voyez, lui dit-il, le visage charmant,
> Et les traits délicats dont la reine est pourvue:
> Je vous jure ma foi que l'accompagnement
> Est d'un tout autre prix; et passe infiniment:
> Ce n'est rien qui ne l'a vue
> Toute nue.
> Je vous la veux montrer sans qu'elle en sache rien.
> (ll. 6-12)

The narcissistic desire of *dédoublement*, the urge to watch an alter ego looking at the queen, poses for the king the inevitable problem of forestalling the desire which will be evoked in Gygès. He must, the king insists, remain unfeeling, must look at the queen as if she were made of marble. The king foolishly believes that Gygès, whom he invites to view the queen "toute nue" (l. 11), will not be aroused by the spectacle and that the queen will not discover the trickery:

> Gygès, il vous faut être sage:
> Point de ridicule désir:
> Je ne prendrais pas de plaisir

Aux voeux impertinents qu'une amour sotte et vaine
Vous ferait faire pour la reine.
Proposez-vous de voir tout ce corps si charmant,
Comme un beau marbre seulement.
(ll. 16-22)

Gygès by gazing on the queen enters a taboo territory, one reserved for her husband alone. By this action Candaule brings upon himself the anger and the revenge of the queen. He deceives himself:

Force gens ont été l'instrument de leur mal;
Candaule en est un témoignage.
(ll. 1-2)

La Fontaine's narrator delights in the irony of the situation—Gygès looking at the queen and experiencing desire for her, while Candaule watches him quite unaware of Gygès's reaction. Further layers of dramatic irony derive from the queen's ignorance of the situation, and then, without Candaule's knowledge, her discovery of the truth and her secretive revenge. So intent is the narrator on transmitting to the reader what he considers to be his own privileged insight into the queen's mind that he suggests on the metanarrative level that the reader should share with him in a voyeuristic literary experience—mirroring the joint experience of Candaule and Gygès in the storyworld. The narrator invites the reader to peer into the narrative situation through an alter ego, through the eyes of a woman:

Je voudrais pour un moment,
Lecteur, que tu fusses femme:
Tu ne saurais autrement
Concevoir jusqu'où la dame
Porta son secret dépit.
(ll. 74-78)

Candaule's exercise in voyeurism once removed consists in replacing Gygès for himself in an action that should be reserved exclusively for the king. Candaule's counterfactual framing of the situation leads him to believe that his dupery will deceive the queen so that she will falsely believe her private space is intact.

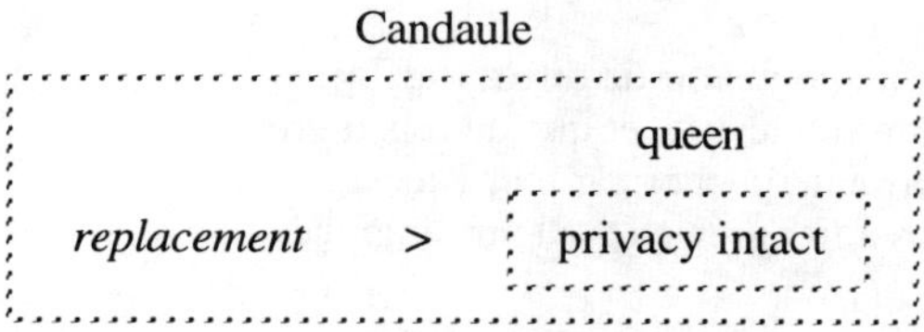

Figure 4.6

In the rapid dénouement the queen operates the ultimate replacement, putting Gygès on the throne and in her bed, thereby directing the reader's gaze toward a juxtaposition of the two acts of deception and introducing a satisfying *trompeur trompé* outcome to the tale:

Car, fût-ce amour, fût-ce rage,
Bientôt la reine le mit
Sur le trône et dans son lit.
(ll. 114-16)

In "La Matrone d'Ephèse"[7] two juxtaposed replacements take place, the first and most significant one is psychological, the second brings about the dénouement. The major deception, however, is self-deception that manifests itself successively in excessive grief, and then in a rapid change in the widow's attitude toward her dead husband. Unlike the narrative development in many other tales, this substitution by replacement does not operate as a factor causing deception, but rather, it reveals a comic element of self-deception. The widow's initial misperception of her own grief as inconsolable is shown to be an error by her swift devaluation of the husband's worth and his replacement by a young soldier as the sole object of her affections.

If self-deception occurs when a person's rational system of dealing with reality is subsumed by an interfering set of feelings or beliefs, this seems to be what is happening in the widow's demonstrations of excessive grief. After her husband's death she had entered into a frenzied mourning:

Celle-ci par ses cris mettait tout en alarme;
Celle-ci faisait un vacarme,
Un bruit, et des regrets à percer tous les coeurs.
(ll. 32-34)

In the narrator's view, her reaction was excessive and self-indulgent:

Chacun fit son devoir de dire à l'affligée
Que tout a sa mesure, et que de tels regrets
    Pourraient pécher par leur excès:
Chacun rendit par là sa douleur rengrégée.
(ll. 39-42)

Motivated by her self-delusional despair, and accompanied by her faithful servant, she enters her husband's tomb and contemplates ending her own life:

Aux consolations la Veuve inaccessible
S'appliquait seulement à tout moyen possible
De suivre le Défunt aux noirs et tristes lieux.
(ll. 63-65)

Meanwhile a soldier is nearby guarding the body of a criminal that is exposed publicly as a deterrent to other robbers. The young soldier knows that if the body disappears he will be punished with death. Nevertheless, his curiosity leads him to investigate the mysterious light emanating from the nearby tomb and to query its occupants with anaphorical persistence:

Il entre, est étonné, demande à cette femme,
    Pourquoi ces cris, pourquoi ces pleurs,
    Pourquoi cette triste musique,
Pourquoi cette maison noire et mélancolique.
(ll. 99-102)

The process of persuading the widow to replace him for her dead husband is undertaken first by her servant's rational pleading:

Qu'importe à votre Epoux que vous cessiez de vivre?
Croyez-vous que lui-même il fût homme à vous suivre
Si par votre trépas vous l'aviez prévenu?
Non, Madame, il voudrait achever sa carrière.
La nôtre sera longue encor si nous voulons.
Se faut-il à vingt ans enfermer dans la bière?
(ll. 127-32)

The soldier's approach is less rational, but equally effective:

Il fait tant que de plaire, et se rend en effet
Plus digne d'être aimé que le mort le mieux fait.
  Il fait tant enfin qu'elle change;
Et toujours par degrés, comme l'on peut penser,
De l'un à l'autre il fait cette femme passer.
(ll. 157-61)

The reader is motivated to ask, especially in light of the ambiguous concluding lines, whether the emotional replacement is not more self-deceptive than was the excessive mourning for her husband. As a cure for her melancholia, finding a new love is a fine adaptive strategy,[8] but the text poses the question of whether her change of heart does not conflict with her own fundamental epistemic principles of loyalty to a departed husband.

A second replacement occurs in the dénouement of "La Matrone d'Ephèse" when the soldier discovers that the body of the criminal has disappeared during his tryst with the bereaved widow. She agrees to allow her husband's body—esteemed earlier as a "trésor" (l. 68)—to replace the criminal's body. Without the final replacement strategy the tale would have ended tragically with the death of the soldier. This substitution by replacement functions in the narrative as an essential element in the dénouement where its purpose is ostensibly to deceive those who would put the soldier to death. Its primary function, however, in the cognitive dimension of the narrative is to reveal further the widow as a victim of self-deception, one who has been brought to such a sorry state that she not only deludes herself into replacing the soldier for her husband in her affections, but now even misjudges the real worth of her husband's body and is willing to consider it as a suitable replacement for the body of a criminal. This is, in any case, the initial misogynist perspective adopted by the narrator:

La Dame y consentit. O volages femelles!
La femme est toujours femme; il en est qui sont belles,
  Il en est qui ne le sont pas.
  S'il en était d'assez fideles,
  Elles auraient assez d'appas.
(ll. 177-81)

The physical substitution of one body for another mirrors the primary replacement, that of the soldier for the husband as an object of love.[9] And in the end, even the narrator has a change of heart when he reduces the whole narrative to an uncomplicated and highly logical replacement schema:

> Car de mettre au patibulaire,
> Le corps d'un mari tant aimé,
> Ce n'était pas peut-être une si grande affaire.
> Cela lui sauvait l'autre; et tout considéré,
> Mieux vaut Goujat debout qu'Empereur enterré.
> (ll. 192-96)

"Les Troqueurs," a tale generally considered to be of La Fontaine's invention, presents a scenario of self-deception generated by a series of replacements. Two ingenious men, Etienne and Gille, believe they have found a way to circumvent the laws concerning indissoluble marriage. Their scheme of legally exchanging wives leads them, however, into a situation where they come to believe falsely that restoration of the original situation would better satisfy their desires. They are the perpetrators of their own delusional space.[10]

The tale is a paradigmatic instance of wishful thinking, which, according to the philosopher Brian P. McLaughlin, is a kind of "unintentional self-misleading" producing "belief not from evidence but in conformity with what one wants to be the case" (44). In the preamble, La Fontaine puts the matter of exchanging wives into the hypothetical mode of wish fulfillment, as the narrator adopts an almost utopian perspective, wondering, albeit ironically, why Rome does not permit the practice:

> Et ne sais pas comment il ne vient de Rome
> Permission de troquer en hymen;
> Non si souvent qu'on en aurait envie,
> Mais tout au moins une fois en sa vie:
> Peut-être un jour nous l'obtiendrons, amen,
> Ainsi soit-il; semblable indult en France
> Viendrait fort bien, j'en réponds, car nos gens
> Sont grands troqueurs, Dieu nous créa changeants.
> (ll. 4-11)

In this *conte* the exchange pattern itself—the *troc*—appears in an apparently uncomplicated form. Etienne and Gille propose the exchange because they have tired of their respective wives, "tous deux étant las / De leurs moitiés" (ll. 18-19). The wives, Jeanne and Tiennette, are not the deceivees. They simply are not part of the decision-making process in a society whose cultural values permit wives, even good-naturedly, to be exchanged much like horses are bartered. The men inquire of a lawyer:

> Vous avez fait sans doute en votre temps
> Plusieurs contrats de diverse nature,
> Ne peut-on point en faire un où les gens
> Troquent de femme ainsi que de monture?
> (ll. 23-26)

As in horse-trading, there is a long bartering session where they weigh the relative merits of Jeanne and Tiennette. Acting on the advice of their lawyer, the husbands, in fact, examine the women in the nude. That scene is certainly a further signal of the fantasyland of wish fulfillment that permeates the tale:

> Or sus, voisins, faisons les choses nettes.
> Vous ne voulez chat en poche donner
> Ni l'un ni l'autre, allons donc confronter
> Vos deux moitiés comme Dieu les a faites.
> (ll. 54-57)

The one whose wife is less beautiful has to hand over a donkey as well. The traditional bartering terms—*top*, *masse*—are exchanged over a drink:

> A toi compère. Et de prendre la tasse,
> Et de trinquer; Allons, sire Oudinet,
> A Jeanne; top; puis à Tiennette; masse.
> Somme qu'enfin la soulte du mulet
> Fut accordée, et voilà marché fait.
> (ll. 73-77)

Jeanne and Tiennette are so pleased with the situation that they consider a similar replacement process: "Bon fait troquer, commère; à ton avis? / Si nous troquions de valet? que t'en semble?" (ll. 101-102) The cognitive frame that is readjusted here by means of the exchange permits the men to consider the enticing possibility of a replacement for their usual sexual partner. The self-deceptive trap they fall into is instigated by their ignorance of the truism that far-off fields look greener. They are lost in a fantasy of anticipatory belief.

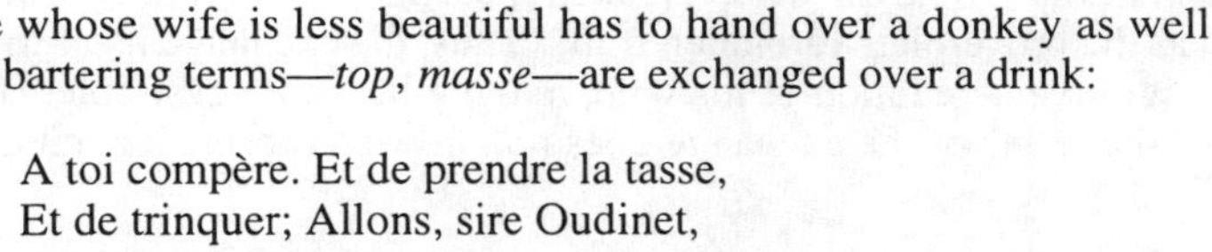

Figure 4.7

As the reader anticipates, both men tire of their new partners. In the dénouement Etienne calls into question the whole legal contract after he unexpectedly finds his wife Tiennette asleep under a tree and makes love with her:

> Qu'en advint-il? Un jour, parmi les bois,
> Etienne vit toute fine seulette
> Près d'un ruisseau sa défunte Tiennette,
> Qui par hasard dormait sous la coudrette.
> Il s'approcha l'éveillant en sursaut.
> Elle du troc ne se souvint pour l'heure;
> Dont le galant sans plus longue demeure
> En vint au point. Bref ils firent le saut.
>
> (ll. 112-19)

The collapse of Etienne's self-deception—he is not happier with his new partner—should lead him to realize that his attraction to Jeanne was stimulated by her unattainability. His own wife, Tiennette, seems more exciting when she is perceived as the proverbial "forbidden fruit." Thus, the exchange itself triggered a different cognitive perception of his wife, making her more desirable. His transgression of the terms of the exchange does not, however, bring a cognitive revelation:

> Le conte dit qu'il la trouva meilleure
> Qu'au premier jour: pourquoi cela? pourquoi?
>
> (ll. 120-21)

Instead, Etienne repeats his mistake by deciding to act on his new cognitive framing of Tiennette. He now falsely believes that by cancelling the exchange, his desires will be satisfied in marriage to his own wife. His self-deception in the form of wishful thinking maintains control of his perception even in the face of counterevidence. Until discovery or real self-awareness sets in, the pattern of exchange could go on indefinitely in a self-repeating circular structure with the unattainable woman always the object of his desire. Forbidden fruit—in this case stolen bread is the metaphor—is much tastier:

> . . . en l'amoureuse loi
> Pain qu'on dérobe et qu'on mange en cachette
> Vaut mieux que pain qu'on cuit ou qu'on achète:
> Je m'en rapporte aux plus savants que moi.
>
> (ll. 122-25)

LOVE AND SACRIFICIAL REPLACEMENT

The next two tales exhibiting a deceit structure based on replacement are "Le Faucon" and "La Courtisane amoureuse" in both of which the lovers are adept at the exchange process. Ironically, only when they believe they have abandoned the exchange system does their desperate recourse to a replacement strategy unintentionally provide such a proof of love that it prompts in the loved one a desire to reciprocate with love. The element of self-deception resides in their persistent tunnel-vision approach to love and in their unawareness that the final gesture of replacement is precisely what will bring about a response of love. For Banville these two tales redeem the rest of the *Contes*:

> Si les contes n'avaient pas gagné leur procès à force de génie et à force de joie, il faudrait leur pardonner encore pour *Le Faucon* et pour *La Courtisane amoureuse*, deux histoires d'amour qu'on relira tant que les langues humaines existeront, et tant que l'amour sera le supplice et la félicité des mortels. Si quelqu'un sait des sacrifices plus attendrissants que le sacrifice de Fédéric et l'humiliation de Constance, si quelqu'un sait de plus beaux discours que le discours de Constance à Camille et que celui de Fédéric à Clitie, que celui-là mette le feu aux *Contes* de La Fontaine et nous n'aurons rien à regretter. (Banville 325)

"Le Faucon"[11] is often considered the most lyric and romantic of the *contes*. In this plaintive tale of unreciprocated love, all of Fédéric's attempts to seduce Clitie, a young married woman, by exchanging money and other gifts for sex—even when she is in desperate straits—meet with her refusal. The would-be lover, after six years of trying to entice her with gifts is himself poor, but still desperately in love. In a brusque nominal style the text enumerates his meager possessions:

> Notre amoureux subsista comme il put
> Dans sa retraite; où le pauvre homme n'eut
> Pour le servir qu'une vieille édentée;
> Cuisine froide et fort peu fréquentée;
> A l'écurie un cheval assez bon,
> Mais non pas fin: sur la perche, un faucon.
>
> (ll. 88-93)

The exchange method of dealing with her has been an utter failure; she remains faithful. A reasonable response would be to shift one's perspective, convinced that there is no evidence to sustain any hope for the future development of a mutual love. In self-deception, however, a typical response is a kind of "inertia of belief," a reluctance to reconstruct one's belief frame even in the face of contradictory data (Rorty 19). Fédéric may be an exemplary representation of true love, but he is also an example of the self-deceptive force of love. La Fontaine's tale, unlike Boccaccio's, is tinged with a skeptical perspective on love.[12] For Jürgen Grimm, Fédéric's presentation as a foolish lover makes of him "un protagoniste négatif, aliéné par l'amour, un 'malheureux amant,' inguérissable et qui manque de sagesse" (223).

The situation changes when misfortune overtakes Clitie. Her husband dies, and then her only child becomes sick. The child wants one thing, the lover's prize falcon. Without it, he will die. Fédéric, however, would never exchange his falcon: "Son maître n'eût donné pour un trésor / Un tel faucon" (ll. 139-40). The text aptly renders Clitie's inner recognition of her situation and her struggle through the style of free indirect discourse:

> . . . Aller ôter encor
> À Fédéric l'unique et seule chose
> Qui lui restait! et supposé qu'elle ose
> Lui demander ce qu'il a pour tout bien,
> Auprès de lui méritait-elle rien?
> Elle l'avait payé d'ingratitude:
> Point de faveurs; toujours hautaine et rude
> En son endroit. De quel front s'en aller
> Après cela le voir et lui parler,
> Ayant été cause de sa ruine?
> D'autre côté l'enfant s'en va mourir;
> Refuse tout; tient tout pour médecine:
> Afin qu'il mange il faut l'entretenir
> De ce faucon: il se tourmente, il crie:
> S'il n'a l'oiseau, c'est fait que de sa vie.
> (ll.141-55)

Motivated by maternal love, she finally goes to see Fédéric, intent on bartering sex for the falcon. He, by now, is so poor that he has nothing to offer her to eat: "Je n'ai, dit-il, cuisinier ni marmite: / Que vous donner?" (ll. 174-75). In a movement of irrational desperation he kills his precious falcon:

Le pauvre amant en ce besoin extrême,
Voit son faucon, sans raisonner le prend,
Lui tord le cou, le plume, le fricasse,
Et l'assaisonne, et court de place en place.
(ll. 179-82)

This desperate act functions in an unforeseen way constructing a simple frame of deceit which radically changes Clitie's perception of Fédéric, as it dispels the doubtful frame into which she had consigned her would-be lover. The reader occupies a privileged cognitive space (knowing the fate of the falcon) as Clitie indulges in a long elegiac speech (after dinner) requesting the bird for her son:

Vous m'aimiez plus que votre propre vie.
A cet amour j'ai très mal répondu:
Et je m'en viens pour comble d'injustice
Vous demander...et quoi? c'est temps perdu;
Votre faucon. Mais non, plutôt périsse
L'enfant, la mère, avec le demeurant,
Que de vous faire un déplaisir si grand.
(ll. 206-12)

Her speech patterns approach the style of high tragedy:

Souffrez sans plus que cette triste mère,
Amant d'amour la chose la plus chère
Que jamais femme au monde puisse avoir,
Un fils unique, une unique espérance,
S'en vienne au moins s'acquitter du devoir
De la nature; et pour toute allégeance
En votre sein décharge sa douleur.
(ll. 213-19)

Fédéric replies in anguished tones:

Hélas! reprit l'amant infortuné,
L'oiseau n'est plus; vous en avez dîné.
L'oiseau n'est plus! dit la veuve confuse.
Non, reprit-il, plût au Ciel vous avoir
Servi mon coeur, et qu'il eût pris la place
De ce faucon: mais le sort me fait voir

Qu'il ne sera jamais en mon pouvoir
De mériter de vous aucune grâce.
(ll. 223-30)

By covertly offering his falcon in the guise of other food he, ironically, wins her love by destroying the one real bartering point he had. By anticipation and without understanding his part in the exchange, Fédéric has effected a felicitous replacement. A program of simple deceit has had unanticipated results. He does not know that he has subverted the exchange program planned by Clitie and substituted by replacement another exchange: the sacrifice of his cherished falcon becomes a proof of love that prompts her reciprocal response. Having received the gift of the falcon as a sign of love, she proffers her hand, a reciprocal sign eagerly accepted by Fédéric:

Elle partit, non sans lui présenter
Une main blanche; unique témoignage
Qu'Amour avait amolli ce courage.
Le pauvre amant prit la main, la baisa,
Et de ses pleurs quelque temps l'arrosa.
(ll. 246-50)

"La Courtisane amoureuse,"[13] is, as Jean-Pierre Collinet concurs, a companion piece of "Le Faucon":

> S'il passe dans les *Contes* de La Fontaine un peu de passion sincère et d'amour vrai, c'est essentiellement grâce à Fédéric et à Constance. La riche fille de joie sert de réplique féminine au cavalier mélancolique et désargenté, formant avec lui contrepoint. D'une nouvelle à l'autre se poursuit un même apprentissage fervent de la tendresse, et de la douceur qu'on trouve à s'oublier soi-même pour l'être aimé. (Collinet 1429)

The replacement effected by Constance constitutes a proof of love and prompts, as it did for Fédéric, a reciprocal response. Constance, a woman whose life has been spent exchanging sex for money, is extremely proud and is not satisfied with ordinary men. Even monsignors are beneath her contempt:

Rome, c'était le lieu de son négoce.
Mettre à ses pieds la mitre avec la crosse
C'était trop peu: les simples monseigneurs

N'étaient d'un rang digne de ses faveurs.
Il lui fallait un homme du conclave;
Et des premiers, et qui fût son esclave;
Et même encore il y profitait peu,
A moins que d'être un cardinal neveu.
Le pape enfin, s'il se fût piqué d'elle,
N'aurait été trop bon pour la donzelle.
(ll. 31-40)

Her nemesis comes when she falls in love with Camille. Not knowing how to express her love, she first of all abandons her system of exchange:

Je vois fort bien que quoi que je vous dise
Je vous déplais. Mon zèle me nuira.
Mais nuise ou non, Constance vous adore :
Méprisez-la, chassez-la, battez-la;
Si vous pouvez, faites-lui pis encore;
Elle est à vous. . . .
(ll. 115-20)

The self-abasement of Constance is juxtaposed with the cruelty of Camille, who takes a perverse pleasure in humiliating her:

Pour en avoir un plus certain indice,
Et s'égayer, et voir si ce coeur fier
Jusques au bout pourrait s'humilier,
Il fit le froid. . . .
(ll. 100-3)

Through her intense love for him she, nevertheless, induces in herself a self-deceptive frame of mind in which even his sadistic games are relegated to the periphery of her attention.[14] He puts her through a series of demeaning experiences, inviting her to act as his lackey as she undresses him and takes off his shoes. Seen from the more rational perspective of the reader, there are elements of excessive self-delusion in her tunnel-vision approach to love, an approach which brings her to the point of being misled, albeit unintentionally, by herself. In fact, the reader expresses surprise at the extent of her abjection and engages in a debate with the narrator:

Ce ne fut tout; elle le déchaussa.
Quoi de sa main! quoi Constance elle-même!
Qui fut-ce donc? Est-ce trop que cela?
Je voudrais bien déchausser ce que j'aime.
(ll. 174-77)

All of Constance's gestures are voluntary and seen by her as necessary replacements for her proud demeanor if she is to win Camille's love. John C. Lapp finds that "her self-abasement has nothing degrading about it" (153); while that view is, I believe, exaggerated, La Fontaine's heroine maintains her dignity by completely controlling the situation. When Camille refuses to help her out of her clothes, she slashes her precious and intricately embroidered clothing with his sword. She eventually expects nothing in return, and is simply willing to sleep at his feet. She kisses Camille's feet—(Does the narrator have a foot fetish? None of these details about feet are in Brusoni, including the earlier removal of Camille's shoes)—and he uses her breast as a cushion for his feet:

Camille donc s'étend; et sur un sein
Pour qui l'ivoire aurait eu de l'envie;
Pose ses pieds, et sans cérémonie
Il s'accommode et se fait un coussin.
(ll. 233-36)

In a nonbartering situation she is no longer powerful. Her substitution of submissiveness for pride stimulates once more a sadistic pleasure in Camille:

On peut juger si Camille était aise.
Quelle victoire! Avoir mis à ce point
Une beauté si superbe et si fière!
(ll. 224-26)

Constance's tactics are viewed by her as a means of altering her lover's framing of their relationship; ironically, her simple "deceit" ends up, I believe, merely in self-deception. Only when she has been reduced to tears, and finally put in her place, does he, near the end of the tale, make a declaration of love and suggest as a further gesture of reciprocity that she in turn be rigorous in her demands of him:

Je suis content, dit-il, de votre amour:
Venez, venez, Constance, c'est mon tour.

Elle se glisse; et lui s'approchant d'elle:
M'avez-vous cru si dur et si brutal
Que d'avoir fait tout de bon le sévère?
Dit-il d'abord, vous me connaissez mal:
Je vous voulais donner lieu de me plaire.
Or bien, je sais le fond de votre coeur.
Je suis content, satisfait, plein de joie,
Comblé d'amour: et que votre rigueur
Si bon lui semble à son tour se déploie:
Elle le peut: usez-en librement.
(ll. 242-53)

But (and here I part ways with Banville and Collinet in their admiration for this love story) Camille's avowed cognitive framing of Constance's past is suspect, for he says that he forgets her past and yet, at the same time, he demands a secret marriage and expects sexual favors before the ceremony! It appears that, consciously or not, he still sees her as a *courtisane*:

Je me déclare aujourd'hui votre amant,
Et votre époux; et ne sais nulle dame,
De quelque rang et beauté que ce soit,
Qui vous valût pour maîtresse et pour femme;
Car le passé rappeler ne se doit
Entre nous deux. Une chose ai-je à dire:
C'est qu'en secret il nous faut marier.
. . . .
L'histoire dit que le drôle ajouta:
Voulez-vous pas, en attendant le prêtre,
A votre amant vous fier aujourd'hui?
(ll. 254-60, 266-68)

The cognitive closure is completed by the narrator's focus on Constance in the concluding lines. Her cognitive framing of the past contrasts with Camille's; for her the prior sexual experiences are nonexistent because love has transformed sexual intimacy into a totally new experience:

Ce que possible on ne croira pas vrai,
C'est que Camille en caressant la belle,
Des dons d'Amour lui fit goûter l'essai.
L'essai? je faux: Constance en était-elle

Aux éléments? Oui, Constance en était
Aux éléments: ce que la belle avait
Pris et donné de plaisirs en sa vie
Compter pour rien jusqu'alors se devait:
Pourquoi cela? Quiconque aime le die.
(ll. 289-97)

Is the tale to be admired unreservedly for its representation of the self-sacrificing love of Constance? In my view, the beautiful depiction of love is sullied because Camille resorts to manipulative tactics and does not share with Constance the self-deceptive myopia of love. While this tale does not present a topographical replacement, its place is here with "Le Faucon"; both tales deal with a sacrificial replacement motivated by love.

***

Gérard Genette has suggested that space in literature functions as more than a simple descriptive element. He describes it as:

> Quelque chose comme une spatialité active et non passive, signifiante et non signifiée, propre à la littérature, spécifique à la littérature, une spatialité représentative et non représentée." (44).

Part of the narrative game in the *Contes* is the constant shifting of characters from legitimate to illegitimate space, the playful forays of husbands, wives, and lovers into forbidden territories. In certain tales, those where deceit operates by displacement, misplacement, and replacement strategies, the dislocating spaces themselves—the *noir séjour* of "Les Quiproquos," the *chambre noire* of "Richard Minutolo," or the dark places of self-delusion and shifting loyalties in "La Matrone d'Ephèse" or "Les Troqueurs"—become the chief builders of another kind of space in the cognitive dimension. They are the tools that, whether they are manipulated intentionally or not, communicate false signals of identification and impede the circulation of knowledge as the characters move across the checkerboard of the text interchanging places, reconfiguring their relationships, and confronting each other with dubious intent. The spaces function as operators that change the interpretive frame, generating counterfactual spaces for the characters and pleasantly ironic perspectives for the readers.

"LE TABLEAU": THE GAME OF LINGUISTIC REPLACEMENT

"Nous voilà parvenus au point"
("Le Tableau," l. 153)

The exchange of scabrous narrative material, an illicit literary activity, is carried out in the *Contes*, like the diegetic seductions, through a playfully deceitful process. In the context of the consensual contract between reader and poet, the "deceitful" nature of poetry is analogous in its functions to the diegetic patterns of deceit. Just as on the diegetic level the characters operate deceit through disguise and substitution, the author, by disguising crudity with appropriately chosen words and metaphors, and by substituting and exchanging veiled allusions for scabrous realities, dupes the willing reader into enjoying "textual fulfillment" (Vincent 55), or as the 1666 Preface expresses it: "au principal point, qui est d'attacher le lecteur, de le réjouir, d'attirer malgré lui son attention, de lui plaire enfin" (Collinet 603).

The game of literary seduction inherent in the *conte en vers* finds iconic expression in "Le Tableau,"[15] the final tale, published by La Fontaine in his last collection of tales, the *Nouveaux Contes*. For Anne L. Birberick, in this tale "the figure of the nun emerges as a figure of the text" (198). Michael Vincent has developed a rich analysis (91-108) of how La Fontaine in "Le Tableau" represents the text metaphorically as a veil that reveals as well as it hides:

> The figure of the veil, however much it owes to figures of vision as model of all modes of perception, resists recuperation and naturalization as merely or even primarily visual by insisting on the partialness, the fragmentary nature of what is perceived across the veil... There always remains a textual residue, a play of signifiers that inhibit complete recuperation of the linguistic and specifically the textual by the visual. Thus the veil begun as a visual metaphor becomes the figure of the text. (102)

La Fontaine explicitly formulates in the introductory lines a linguistic and poetic theory grounded in a polysemic view of language:

> Quand le mot est bien trouvé,
> Le sexe, en sa faveur, à la chose pardonne;
> Ce n'est plus elle alors, c'est elle encor pourtant.
> (ll. 25-27)

For La Fontaine, the well-chosen word is replete with connotations that invest the text with what one of La Fontaine's editors, Alain Bassy, terms a third dimension:

> La langue des *Contes* n'est jamais simple: elle est chargée de connotations multiples; elle ne signifie que dans la référence; elle est toujours renvoyée à l'extérieur d'elle-même. C'est par là qu'elle acquiert une troisième dimension et qu'elle mobilise l'entendement (*intellectus*); cette troisième dimension du langage, le poète le trouve dans l'anachronie ou dans l'équivoque du sens... Dans chaque mot, dans chaque expression, le poète laisse, volontairement, se déposer et se recouvrir des niveaux de signification différents... Il n'existe pas une lecture possible des *Contes*, mais une multiplicité de lectures imaginables, dont l'auteur a volontairement semé les germes. Ecriture piégée, les *Contes* induisent le lecteur vers des significations qu'ils feignent de déjouer ensuite. (17-18)

Michel Picard in *La Lecture comme jeu* judiciously links the ambivalence of verbal game playing with the process of subversion and with the ludic character of reading itself:

> On notera enfin que le critère de la polysémie, si clairement lié à la nature ludique de la lecture littéraire, se caractérise par la même ambivalence que la subversion—ce qui ne surprendra pas, vu leurs profondes relations. (259)

La Fontaine's challenge in "Le Tableau," as it is to a greater or lesser extent in each of the *contes*, is to present an improper subject in an amusing, but not offensive manner, to his readers:

> On m'engage à conter d'une manière honnête
> Le sujet d'un de ces tableaux
> Sur lesquels on met des rideaux.
> (ll. 1-3)

The narrative program of presenting a taboo subject—the erotic scene—but disguising it through a kind of literary "deceit," is enigmatically alluded to in lines 35-36:

Je veux, quoi qu'il en soit, expliquer à des belles
Cette chaise rompue, et ce rustre tombé.

The focus of the tale, from the outset, is displaced from the events of the narrated tale itself to the unfolding of the narrator's technique, his "manière de conter" (1665 Preface). The reader wonders how the narrator will fulfill his promise:

Puisqu'on le veut ainsi, je ferai de mon mieux:
Nuls traits à découvert n'auront ici de place.
Tout y sera voilé, mais de gaze; et si bien,
Que je crois qu'on n'en perdra rien.
(ll. 18-21)

The challenge for the narrator is to find words "Qui disent et ne disent pas" (l. 6)—the gauze or veil that reveals and hides at the same time. The *mot* well-chosen makes the *chose* more acceptable: "Ce n'est plus elle alors, c'est elle encor pourtant" (l. 27). Patterns of substitution, exchange, and disguise are operating on the metanarrative level of the text in "Le Tableau."

The narrative begins at line 45, but the dramatic action and the depiction of the taboo subject is constantly delayed. The narrator teases the reader, withholding textual fulfillment (to use again Vincent's apt term). Just as the nuns are in *attente*, waiting for "le bachelier trop lent," the reader must wait as the narrator indulges in all the textual foreplay of describing the nuns, the flowers, the wine, and the embroidered tablecloth of their opulent dining room, which will never be used in the ensuing tale. The reader, as well as the nuns, awaits "le bachelier trop lent" (ll. 106-14).

On the diegetic level, first the replacement of a stereotypical young fool for the too-long-awaited lover (ll. 115-40), and then the two nuns' taking turns making love with him, are figures of another exchange process, a linguistic substitution being activated on the poetic level in the climactic passage. A juxtaposition pattern operating in this tale opposes within the reader-observer's cognitive frame a replacement on the diegetic level and a replacement on the metanarrative level.

In his narration of the erotic scene in which the *sot* falls off his chair, La Fontaine builds up an erotically-charged semantic network of words related to *chaise* (l. 160), *cathédral* (l. 172), *tribune* (l. 177), *trône* (l. 187), *poste* (l. 189), *occupé* (l. 193), *débusquée* (l. 197), *détrônée* (l. 229).[16] Overtly, he is speaking about a chair, but the change in connotation starts taking place with the expressions, "galant cathédral" and "amoureuse tribune," closely followed by the even more explicitly sexual metaphors, *tramontane* and *timon*:[17]

Thérèse en ce malheur perdit la tramontane,
Claude la débusqua, s'emparant du timon.
    Thérèse pire qu'un démon
Tâche à la retirer, et se remettre au trône;
    Mais celle-ci n'est pas personne
    A céder un poste si doux.
(ll. 184-89)

Intertwined with the metaphoric field of the chair is the metaphor of the *discours—ex cathedra*—from the cathedral chair, the "amoureuse tribune." Soeur Thérèse is the orator; her "speech," replete with rhetorical *action* (l. 175), *émotion* (l. 176), and a *point* leading to an unfortunate *conclusion* (l. 180), is an extended metaphor of her sexual activity:

Ses jeux sont violents. A terre on vit bientôt
Le galant cathédral; ou soit par le défaut
De la chaise un peu faible; ou soit que du pitaud
    Le corps ne fût pas fait de plume;
Ou soit que soeur Thérèse eût chargé d'action
Son discours véhément, et plein d'émotion;
On entendit craquer l'amoureuse tribune.
Le rustre tombe à terre en cette occasion.
    Ce premier point eut par fortune
    Malheureuse conclusion.
(ll. 171-80)

Recalling the erotic connotations of the word *point* (Guiraud: "*point*-coït") and considering the context of the broken chair, another meaning may also enter into play here (Guiraud: "*point (être en)*-être en érection"). The word occurs earlier with a sexual connotation in line 136: "Ce pitaud doit valoir pour le point souhaité." Significantly, on the metanarrative level, the narrator in line 153 refers to his arrival—with the reader—at the climactic part of the tale in terms that have an alternative erotic meaning (Guiraud: "*venir au point*-coïter"). Reader and narrator are involved intimately in a textual act: "Nous voilà parvenus au point."

The narrator has successfully met the challenge of fabricating a linguistic veil for the licentious scene, further tempering the barely occluded obscenity along the way with epic alexandrines, mythological allusions (ll. 200-221), and lyric invocations, first to the "Dieu des vers" (ll. 154-60), next to the censors (l. 181), and then to the "gens de bien" (ll. 182-83). As "Le Tableau" illustrates, the *conte en vers* genre has as a distinctive characteristic the juxtaposition of crude subject matter with an elegant, albeit "neglected," mode of expression. The reader's pleasure derives to a large extent from an awareness of the *décalage* between the elevation of poetry and the *bassesse* of the narrative material. One might say that in the seduction of the reader, too, part of the pleasure derives from the broken chair; the poetry is irregular. La Fontaine alludes in the 1666 Preface to "ces sortes de négligences qu'il ne se pardonnerait pas lui-même en un autre genre de poésie; mais qui sont inséparables, pour ainsi dire, de celui-ci" (Collinet 603). As La Fontaine affirmed (Collinet 556), licentious subject matter is an essential component of his *conte en vers* and it is precisely the sexual *duperie* that offers a rich field in which to cultivate a poetics grounded in the cognitive space of the text.

Like the game of love, the seduction of the reader is in its own way a game. The recurring motif of love as a *jeu* finds a metadiegetic transposition in the 1665 Preface where La Fontaine, in an attempt to ward off the censors, describes his writing of the *Contes* as *jeu*:[18] "Mais qui ne voit que ceci est jeu, et par conséquent ne peut porter coup?" (Collinet 537). Using the resources of poetry and narrative, the *Contes* lure the reader into a delusory space where society's taboos cease to function.

The narratives of seduction in the storyworld become, in a sense, a metaphoric mirror of the poetic and narrative situation.[19] La Fontaine's implicit reader is willingly seduced into the world of poetry and accepts a pact with the author to share "forbidden" tales, *contes* judged by the censors as unacceptable for the genteel public,[20] and generally considered by the literary establishment as unworthy of serious consideration.

NOTES

1. According to Guiraud's *Dictionnaire érotique*: "*en venir au point*-coïter."

2. See R. W. Mitchell and N. S. Thompson, eds. *Deception: Perspectives on Human and Nonhuman Deceit.*

3. "L'Anneau d'Hans Carvel" also belongs in the category of misplacement. A jealous husband dreams that a devil places a ring on his finger. The devil assures him that so long as he does not remove this ring he will not be a cuckold. When he awakes beside his wife he discovers that he has "le doigt où vous savez."

4. For a comparison of Boccaccio's version with three fabliaux and with Chaucer's "Reeve's Tale," see Almansi, 66-69.

5. See *Perspectives on Self-Deception*, eds. Brian P. McLaughlin and Amélie Oksenberg Rorty. On the so-called paradoxes of self-deception, see Herbert Fingarette, *Self-Deception*, chapter 1. A skeptical viewpoint is expressed in M. R. Haight, *A Study of Self—Deception*.

6. See Collinet (1471) for seventeenth-century analogues; more recently the tale was the basis of André Gide's play, *Le Roi Candaule*.

7. First published at the end of *Le Poème du Quinquina* (1682). See Collinet (1306-07) for a detailed history of the many versions of the *Matrone d'Ephèse* story from Petronius to Voltaire.

8. Amélie Oksenberg Rorty sees self-deception as "an effective, if irrational cure for melancholia"—see "The Deceptive Self: Liars, Layers, and Lairs," 17.

9. Another dénouement in which replacement occurs is, of course, "Les Lunettes" where—just as in "La Matrone d'Ephèse"—it is the means by which the young man escapes punishment. The young man who had disguised himself as a nun is tied to a tree by the older nuns who go off to get whips with which to beat him. He cleverly escapes by convincing a miller to take his place. He tricks the miller into the false belief that the nuns have tied him up in order to have sexual relations with him. The replacement does not deceive the nuns, nor is it intended to deceive them. As far as they are concerned the replacement is satisfactory:

> Quoi! tu n'es pas notre faiseur d'enfants?
> Tant pis pour toi, tu paieras pour le sire:
> Nous n'avons pas telles armes en main,
> Pour demeurer en un si beau chemin.
>
> (ll. 179-82)

The substitution by replacement is a narrative expedient, but again shows up the foolishness of those who accept its validity.

10. While the exchange to which they consent is a verbal agreement witnessed by a notary, I am considering this tale as being in the category of nonverbal deception because what changes the perception and leads to the delusion is not the moment of verbal agreement but the actual experience of substituting wives. Similar at first sight to "Les Troqueurs" is "Pâté d'Anguille" where a man tires of his wife and wants to sleep with his valet's wife. The deceit itself is, however, based not on replacement, nor on self-delusion, but is verbal and based on a clever abuse of metaphor.

11. See Boccaccio, *Decameron* V, 9, and two structuralist analyses, one of the Boccaccio tale (Todorov 73-75) and the other a comparison of La Fontaine and Boccaccio's tales (Genot, "Le récit [du] déclassé").

12. See John Lapp on the skepticism of this tale (*Esthetics* 63) and Donna Kuizenga, "La Fontaine's 'Le Faucon': A Lesson of Experience."

13. French critics resisted for some time acknowledging Girolamo Brusoni's *Curiosissime novelle amorose* as the possible source of this tale, perhaps because the Gaston Paris article on "La Courtisane amoureuse" was published in Italy, not in France. Vittorio Lugli's article, "La Courtisane amoureuse," presents the saga of this strange example of French insularity: "Gli editori francesi continuano ad ignorare lo scritto del Paris; il racconto non avrebbe 'aucune source précise'" (Lugli 720). [The French editors continue to be unaware of the article by Gaston Paris; the tale, according to them has "aucune source précise"—my translation.] Neither the article published in 1901 by Gaston Paris in an Italian *festschrift* volume nor Lugli's own book *Il prodigio di La Fontaine* in 1939, his "La Fontaine e il glottologo" published in 1953, nor his *Tre mezzi secoli* (1955) resulted in any reference to Brusoni in French editions of the *Contes*. Not until the 1980 edition of the *Contes* by Collinet and Ferrier did the statement appear: "Gaston Paris a rapproché le conte de La Fontaine de 'La Cortegiana innamorata' dans les *Curiosissime novelle amorose* de Girolamo Brusoni" (434). Brusoni's tale, difficult to find, is reproduced in the Gaston Paris article (378-80).

14. Rorty again argues that such tunnel vision, when it is prompted by strong emotion, constitutes a bona fide case of self-deception: "But when a person is afraid, or absorbed in love or grief… she can fail to integrate the relevant material that is at the periphery of her strong, attentive focusing" (19).

15. The source is a tale told on the first day in Aretino, *Ragionamenti: Sei Giornate* (29-32), but, apart from the basic elements of the story, La Fontaine's text bears little resemblance to Aretino's.

16. See in Guiraud, *Dictionnaire érotique*: "*trône*-sexe de la femme"; and religious terms used metaphorically in a similar sense: *chapelle*, *temple*, *tabernacle*; *poste* likewise designated the female sex organ. La Fontaine's *trône*, *galant cathédral* and *amoureuse tribune*, *poste*, play on these erotic connotations to suggest the male sex organ. The word *débusquée*, according to Guiraud—"faire sortir le pénis"; La Fontaine uses *détrônée* with the same signification.

17. See Guiraud, "*timon*-pénis," followed by the quotation of La Fontaine's line 185 from "Le Tableau."

18. See in Tyler, *Concordance* the numerous references to the "jeu d'amour" in the text of the *Contes*.

19. Ross Chambers finds in nineteenth-century French fiction a similar metaphor of narrative seduction, an attempt to "mobilize forces of desire (desire to narrate and desire of narrating) in ways that lie outside the cultural norm of, let us say, 'legitimate' exchange" (10).

20. As for the female readers whom the censors would like to protect, La Fontaine's attitude toward their reading of the *Contes* ("J'ouvre l'esprit, et rend le sexe habile / A se garder de ces pièges divers"—[Collinet 903]) is comparable to the narrator's statement about diegetic seduction: "Heureux sont ceux qu'on trompe à leur profit" ("La Mandragore" l. 288).

CHAPTER FIVE

LIES AND THE RHETORIC OF SEDUCTION

Je ne connais rhéteur, ni maître ès arts.
Tel que l'Amour; il excelle en bien dire.
"*La Confidente sans le savoir*
ou le Stratagème," ll. 1-2

Lying is the transgression of a basic contract of language.[1] The implicit contract is that a speaker is morally bound to use words that are intended to convey his or her thought. On the other side of the contract in a language situation, the listener agrees to believe the words that are spoken. When the speaker transgresses this fiduciary contract by misrepresenting thoughts or feelings, a lie is produced. In the context of my analysis of the cognitive structures of deceit, a lie is a space-builder whose aim is to shift the hearer's interpretative framing of a given situation into the counterfactual mode. In a scenario of seduction the false information generates the cognitive frame within which the one to be seduced makes a decision to consent to the sexual overtures. Lies also serve other purposes in a variety of sexual contexts. At times they preserve the secret of an illicit relationship; at others, they facilitate lovers' encounters through misleading signals.

Lies in La Fontaine's *Contes* are not seen in moral terms. They are simply clever strategies functioning in a playfully amoral universe. Philosophers and theologians have, however, developed over the centuries classifications for lies that prove useful for grouping La Fontaine's tales to analyze their patterns of deceit.[2] The consideration of intentionality is absolutely essential to the analysis and categorization of lies. As Marcel Eck states: "The essential criterion of lying remains the same in all cases: namely, the *intention* that motivates it" (25). A lie is usually told in order either to gain some benefit or to protect oneself, the former cognitive program of the deceiver being offensive and the latter, defensive. In order to be elected a politician claims falsely to have no underworld connections. A man states insincerely that he loves a woman so that she will go to bed with him. Both the politician and the lover are indulging in offensive—as opposed to defensive—lies that create a false belief space in the listener. In the defensive lie, the aim is usually to prevent an action from occurring by changing the interpretation of a situation in the mind of the deceivee. A student accused of cheating on an exam claims not to have done so. That is a defensive lie. Lies are morally subversive not only in their contravention of the fiduciary contract, but likewise in that they induce the listener to make faulty and uninformed choices. The consequence of a successful lie is the hearer's decision to act, or not to act, based on false information.

THE OFFENSIVE LIE

An offensive lie occurs sometimes in a minor role at the beginning of a narrative when proffering preliminary misleading information prepares the way for the main act of deception. In "Richard Minutolo" Richard lies to Catelle, telling her that her husband has a mistress, Simone, whom he is to meet in a dark room at the baths. Richard even insists on his own good intentions in revealing the infidelity of her husband: "Mais, grâce à Dieu, je ne veux rien de vous: / Ce qui me meut n'est du tout que bon zèle" (ll. 53-54). Catelle's consequent false cognitive framing of the situation causes her to decide on a displacement tactic by which she substitutes herself for Simone in the dark room—where she unknowingly makes love to Richard.

In "Le Cocu battu et content," a tale where disguise is the operative space-builder, Messire Bon's wife lies to her husband by informing him that the young falconer has promised to be waiting for her in the garden that night. As a consequence of his counterfactual framing of the situation, Messire Bon, disguised as his wife, mounts a vigil in the garden, while the falconer occupies his bed and his wife.

In "Les Rémois" a woman's initial lie sets the stage for a scene charged with dramatic irony in which two husbands assist, hidden and silent, at the spectacle of their wives' seduction. To lure the men into the trap, the woman invites them to meet her during her husband's supposed absence:

> La dame donc pour tromper ses voisins
> Leur dit un jour: Vous boirez de nos vins
> Ce soir chez nous. Mon mari s'en va faire

Un tour aux champs; et le bon de l'affaire
C'est qu'il ne doit au gîte revenir.
Nous nous pourrons à l'aise entretenir.
(ll. 58-63)

La Fontaine seems to recognize that such a lie with its deliberately destructive consequences—and particularly because it is uttered by a woman—needs some justification. Addressing himself to his female readers, he considers how they are exploited by men and urges them to retaliate in kind:

Vous y servez de jouet et de proie
A jeunes gens indiscrets, scélérats:
C'est bien raison qu'au double on le leur rende:
Le beau premier qui sera dans vos lacs,
Plumez-le-moi, je vous le recommande.
(ll. 53-57)

As the principal means of deceit, the offensive lie is the tactic employed in the story told by the third woman in "La Gageure des trois commères" (ll. 216-333). I am stretching the definition of a lie here to include a nonverbal sign that functions exactly like a lie.[3] The misleading sign is a string by which a clever woman—a "femme d'esprit" (l. 230)—manages to keep her husband, Henriet Berlinguier, away from her bed for three nights while she entertains her lover in his stead. She attaches a string to her toe, then makes sure that her husband feels it as he turns over in bed. Henriet Berlinguier is immediately suspicious and, as he gets up and follows the string out into the street, the reader follows the string of his thought as it develops into a theory that the string is a means by which his wife's lover will signal to her his presence. Led on by his false framing of the situation, he pursues the search:

. . . suit le fil dans la rue:
Conclut de là que l'on le trahissait:
Que quelque amant que la donzelle avait,
Avec ce fil par le pied la tirait,
L'avertissant ainsi de sa venue:
Que la galande aussitôt descendait,
Tandis que lui pauvre mari dormait.
Car autrement pourquoi ce badinage?
(ll. 253-60)

By the time he gives up his vigil in the street and returns to bed, the lover has departed.

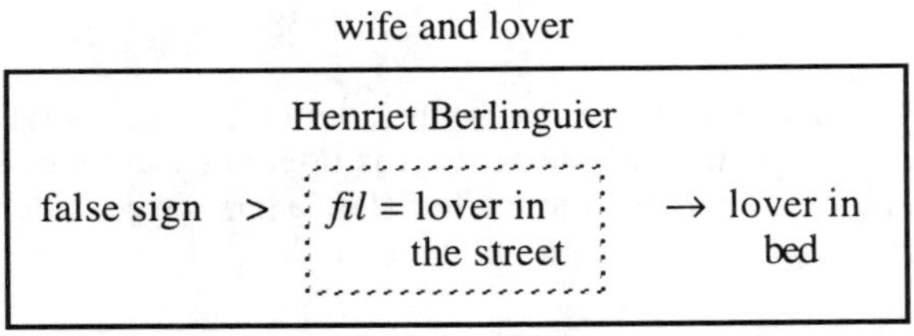

Figure 5.1

On the third night the lovers decide that it is time to bring their duplicitous comedy to its dénouement:

> . . . Leur ardeur refroidie,
> Il en fallut venir au dénouement;
> Trois actes eut sans plus la comédie.
> (ll. 293-95)

Again the "fil" is the misleading sign, but this time it operates to accomplish the cognitive closure of the tale. One of the lover's friends pulls on the string, is captured by Henriet Berlinguier, and in the farcical scene that ensues the husband is made to believe that the string was supposed to be on the toe of the "chambrière." His wife explains how she had seen a string attached to the young woman's toe and had decided to catch the intruder by attaching a similar string to her own toe. The husband revises his interpretive frame and retrospectively sees the whole drama as a comedy whose final scene is typically a happy marriage:

> . . . Berlinguier fut plus doux;
> Dit qu'il fallait au lendemain attendre.
> On les dota l'un et l'autre amplement;
> L'époux, la fille; et le valet, l'amant:
> Puis au moutier le couple s'alla rendre.
> (ll. 317-21)

For the lovers, the scene represents the final dupery of Henriet Berlinguier who never discovers the true scenario. Ironically, his first false interpretation of the "fil" was much closer to the truth.

"La Clochette" also belongs in this category and, as in the tale of the third woman in "La Gageure des trois commères," the trickery operates through a deceptive sign. Even La Fontaine feels uncomfortable about the subject matter of this *conte*; in its preface the narrator states:

> Et, supposé que quant à la matière
> J'eusse failli, du moins pourrais-je pas
> Le réparer par la forme en tout cas?
>
> (ll. 15-17)

A young man spies a 13-year-old shepherdess—"elle était trop jeunette, / Et d'âge encore incapable d'aimer" (ll. 32-33). He takes a bell from one of her sheep and chases the sheep away. When, later on, her mother sends the girl out to find the sheep, the "seducer" rings the bell to lure her far off into the woods where the rape takes place:

> Jugez, lecteur quelle fut sa surprise
> Quand elle ouït la voix de son amant.
> Belle, dit-il, toute chose est permise
> Pour se tirer de l'amoureux tourment.
> A ce discours, la fille toute en transe
> Remplit de cris ces lieux peu fréquentés;
> Nul n'accourut . . . .
>
> (ll. 62-68)

By way of cognitive closure, the narrator, with no condemnation of the young man's act, simply advises (somewhat halfheartedly) female readers to learn by this example:

> . . . O belles évitez
> Le fond des bois et leur vaste silence.
>
> (ll. 68-69)

THE INTRUSIVE LIE

A subcategory of the offensive lie can be termed the "intrusive lie."[4] This is a lie that, somewhat like the Trojan horse, succeeds in invading the territory of the deceivee under a feigned appearance. The lie is transported quite innocently by a third party into a territory that is off-limits to its sender. When the lie arrives before its decoder, he or she recognizes, but not immediately, that it contains a hidden meaning.

In "La Confidente sans le savoir ou le Stratagème" Cléon's prudish guardian, Madame Alis, fulfills the function of the Trojan horse.[5] Aminte manages to declare her love and arrange a rendezvous with Cléon, the man she loves, by falsely stating to Madame Alis that Cléon has been importuning her with letters and other signs of love:

> Sous ma fenêtre il passe incessamment;
> Je ne saurais faire un pas seulement
> Que je ne l'aie aussitôt à mes trousses;
> Lettres, billets pleins de paroles douces,
> Me sont donnés par une dont le nom
> Vous est connu; je la tais pour raison.
> Faites cesser pour Dieu cette poursuite.
> (ll. 66-72)

After two upbraidings by Madame Alis, Cléon finally decodes the covert message that the lies are declarations of love:

> Il s'en retourne, il rumine, il repense,
> Il rêve tant qu'enfin il dit en soi:
> Si c'était là quelque ruse d'Aminte?
> Je trouve, hélas, mon devoir dans sa plainte.
> Elle me dit: O Cléon! aime-moi,
> Aime-moi donc . . . .
> (ll. 113-18)

Aminte even succeeds in getting true information to him about when her husband will be absent—by lying! She does so by complaining again to Alis about how bothersome all Cléon's presents are and informing her that, what is more, he even has stated his intention to visit her that very night when her husband is away:

> Oyez le reste il m'a fait dire aussi
> Qu'il sait fort bien qu'aujourd'hui pour affaire
> Mon mari couche à sa maison des champs;
> Qu'incontinent qu'il croira que mes gens
> Seront couchés, et dans leur premier somme,
> Il se rendra devers mon cabinet.

Qu'espère-t-il? pour qui me prend cet homme?
Un rendez-vous! Est-il fol en effet?
(ll. 149-56)

All this is duly reported to Cléon by the unknowing *entremetteuse*. Cléon, who now fully comprehends the situation, adopts the strategy of lying himself in his reply to Alis:

J'aimais, dit-il, il est vrai, cette belle;
Mais puisqu'il faut ne rien espérer d'elle,
Je me retire, et prendrai ce parti.
(ll. 170-72)

The lies of Aminte in this tale are double-edged and consequently generate an embedded deceit structure. They are intended to deceive Alis, but simultaneously to inform Cléon.

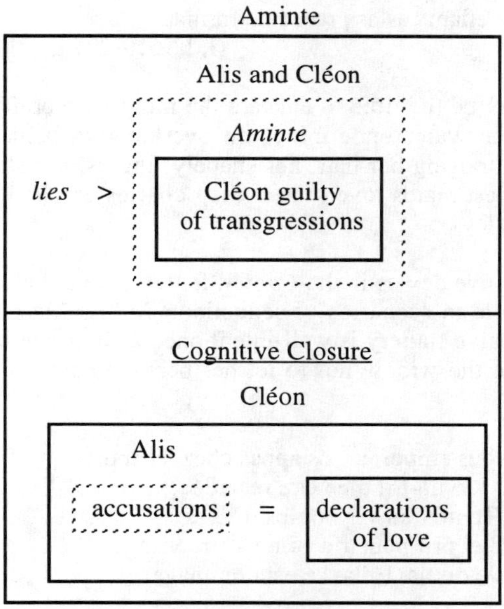

Figure 5.2

In the first half of the above diagram, Alis and Cléon share the same false belief that Aminte believes Cléon to be guilty of certain social transgressions. At the close of the tale Cléon has come to understand, still by means of the lies, that what Alis views as accusations are really declarations of love.

FLATTERY

Flattery is usually considered by moral philosophers as a trivial lie, very much in the same category as the white lie, or what Thomas Aquinas terms the "jocose lie," a lie told with a desire to please and not intended to cause harm (Bok 58). Flattery in its role as an offensive tactic in seduction has long been considered a legitimate part of the lover's panoply of persuasive devices. In Ovid's *Ars amatoria* the lover is urged to use flattery:

> Blanditiis animum furtium deprendere nunc sit
> Ut pendens liquida ripa subitur aqua.
> Nec faciem, nec te pigeat laudare capillo,
> Et teretes digitos exiguumque pedem,
> Delectant etiam castas praeconia formae.
> (I, ll. 619-23)

> (Now be the time to ensnare the mind with crafty flatteries, as the water undermines an overhanging bank. Nor be wary of praising her hair, her shapely fingers, her small foot; even honest maids love to hear their charms extolled. —trans. J.H. Mozley)

Not a major deceptive device in the *Contes*, flattery, nevertheless, plays an effective supporting role in scenarios of seduction.[6] In "La Matrone d'Ephèse" the rhetoric of persuasive flattery is well-intentioned as both the lover and the widow's servant urge the widow not to let her bereavement bring about her own death:

> Voulez-vous emporter vos appas chez les morts?
> Que vous servira-t-il d'en être regardée?
> Tantôt, en voyant les trésors
> Dont le Ciel prit plaisir d'orner votre visage,
> Je disais: Hélas! c'est dommage!

Nous-mêmes nous allons enterrer tout cela.
A ce discours flatteur la Dame s'éveilla.
(ll. 136-42)

The narrator comments on the efficacity of flattery as a means of persuasion in love:

Voilà donc notre Veuve écoutant la louange,
Poison qui de l'amour est le premier degré:
La voilà qui trouve à son gré
Celui qui le lui donne...
(ll. 153-56)

In "L'Oraison de saint Julien" Renaud, playing his role of the ideal lover, begins the seduction of his hostess by praising her beauty:

Renaud poursuit: louant par le menu
Tout ce qu'il voit, tout ce qu'il n'a point vu,
Et qu'il verrait volontiers si la belle
Plus que le droit ne se montrait cruelle.
Pour vous louer comme vous méritez,
Ajouta-t-il, et marquer les beautés
Dont j'ai la vue avec le coeur frappée.
(Car près de vous l'un et l'autre s'ensuit)
Il faut un siècle, et je n'ai qu'une nuit.
(ll. 279-87)

SILENCE

If lying is a means of keeping thoughts, intentions, and feelings secret, another tactic for hiding the truth through secrecy is silence. Silence, like lying, can be part of an offensive strategy for generating false beliefs and facilitating trickery. By feigning to be mute, the seducer in "Mazet de Lamporechio"[7] uses silence as a weapon of deceit. Mazet's ignoble intentions are clear from the outset:

Trop bien croyait, ces soeurs étant peu sages,
Qu'il en pourrait croquer une en passant,
Et puis une autre, et puis toute la troupe.
(ll. 69-71)

The pretense of being mute and stupid tricks the nuns into seeing themselves as the manipulators:

> . . . Dedans ce cabinet
> Menons ce sot: Mazet était beau mâle,
> Et la galande à le considérer
> Avait pris goût; . . . .
> . . . .
> Il est muet et n'en pourra rien dire.
> (ll. 117-20, 141)

Only when Mazet himself tires of the charade does he miraculously shed his self-enforced silence to complain about being overburdened by the nuns' sexual demands:

> J'ai toujours ouï, ce dit-il, qu'un bon coq
> N'en a que sept, au moins qu'on ne me laisse
> Toutes les neuf. Miracle, dit l'abbesse,
> Venez, mes soeurs, nos jeûnes ont tant fait
> Que Mazet parle. A l'entour du muet,
> Non plus muet, toutes huit accoururent.
> (ll. 180-85)

The pattern of deceit resembles the embedded counterfactual spaces in "Le Cocu battu et content" where the disguise was used to make the deceivee falsely believe that he himself was duping someone else by the disguise. Here, silence (instigated by Mazet) functions as a builder of false cognitive spaces in which the nuns believe they are manipulating Mazet by means of his silence. The nuns do not discover that they have been manipulated by Mazet; their only concern at the end of the narrative is to keep Mazet and their secret within the convent walls. (See Figure 5.3 below)

Silence functions as an active operator of deceit in "Les Rémois," where the deceivers lead their victims to believe that the defensive silence into which they are coerced is of their own choosing. The complexity of the deceit pattern is not that of the classic *trompeur trompé* scheme because there is even a further level of cognitive manipulation. The primary "trompeurs" are the husband and wife; the victims falsely believe they are "trompeurs trompés." Having been lured to an amorous rendezvous on a false pretext, the two men hide in a closet when the husband returns, as they believe, unexpectedly.

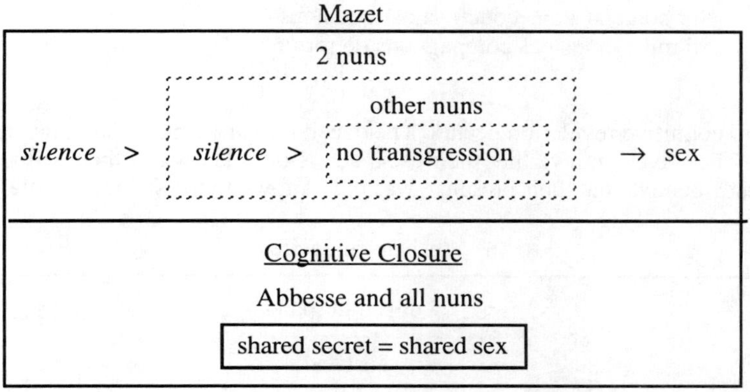

Figure 5.3

The wife tells her husband that the festive dinner she has prepared is for the two neighbors' wives, Alis and Simonette. So far, the hidden spectators are merely amused at the ingenuity of the hostess. As the husband, however, pays ever more attention to their wives, they become increasingly uncomfortable. In a diptych structure, first Simonette, then Alis, is led off into the wine-cellar by the wife, leaving the husband alone to seduce the other wife. The seduction scenes are torturous for the hidden spectators who only keep silent because they believe they have brought the adventure upon themselves. When the first husband, beside himself with rage, is tempted to reveal himself, his companion urges him to keep silent:

> Gardez-vous bien de faire une sottise,
> Lui dit tout bas son compagnon d'amours,
> Tenez-vous coi; le bruit en nulle guise
> N'est bon ici; d'autant plus qu'en vos lacs
> Vous êtes pris: ne vous montrez donc pas.
> (ll. 141-45)

When it comes to the turn of Alis, her husband has to be restrained:

> Nous ne voulons sur vous nul avantage.
> C'est bien raison que messer Cocuage

Sur son état vous couche ainsi que nous.
Sommes-nous pas compagnons de fortune?
(ll. 188-91)

No cognitive revelation occurs. The deceit remains embedded in layers of silence. The wives believe that their escapade is only known to their seducer; the husbands believe that their presence was only known to the seducer's wife.

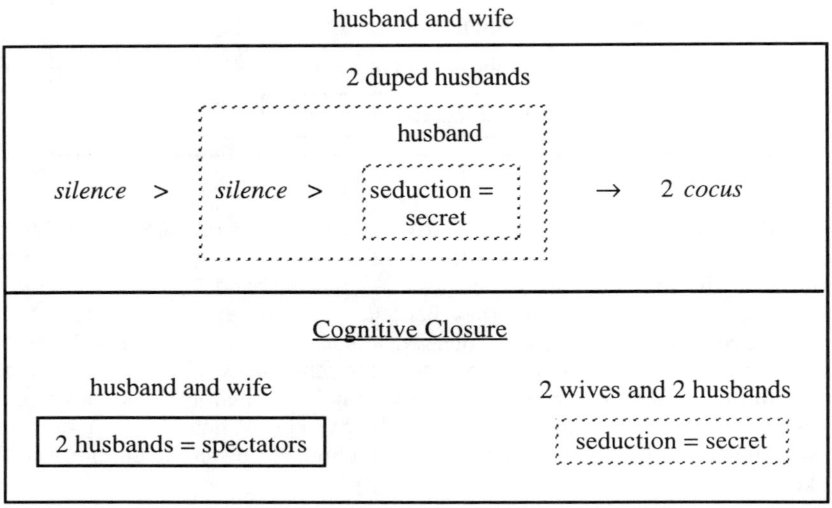

Figure 5.4

THE DEFENSIVE LIE

The defensive lie is based on the simple principle of utilitarianism. Sissela Bok would describe this category of lie as a "lie in a crisis" (Chapter 8). Since its aim is to avoid harmful consequences, it is not generally a tool of seduction.

Two tales whose deceit structure relies on a strategic defensive lie involve duping a wily but naive devil. The devil's knowledge of such practical human concerns as agriculture and sexuality is, so to speak, a blank cognitive space. These are not tales of seduction; the lies play on fear of women and marriage. The use of sexuality to ward off the devil is in itself part of the ironic structure of the text. The first *conte*, "Le Diable de Papefiguière" whose source is Rabelais' *Quart Livre* (XLV-XLVII), depicts a devil determined to take revenge on Phlipot, a poor farmer in the make-believe land of Papefiguière. The devil's lack of *savoir* is attested to by himself (and here La Fontaine aims a satirical jibe at the *gentilhommes* of his own society)—"Je t'ai déjà dit que j'étais gentilhomme, / Né pour chommer, et pour ne rien savoir" (ll. 73-74), and by Perrette—"mon petit doigt saurait plus de malice, / Si je voulais, que n'en sait tout son corps" (ll. 152-53). Perrette's defensive lie, which succeeds in its aim of chasing the devil away, consists in showing him a part of her anatomy he knows nothing about and telling him that Phlipot, having wounded her, is about to attack him:

> Pour s'éprouver le perfide m'a fait
> Cette balafre. A ces mots au follet
> Elle fait voir...Et quoi? Chose terrible.
> Le diable en eut une peur tant terrible
> Qu'il se signa, pensa presque tomber;
> Onc n'avait vu, ne lu, n'ouï conter
> Que coups de griffe eussent semblable forme.
> Bref, aussitôt qu'il aperçut l'énorme
> Solution de continuité,
> Il demeura si fort épouvanté,
> Qu'il prit la fuite et laissa là Perrette.
> (ll. 174-84)

In "Belphégor," the other tale about an unsophisticated devil, the only way to trick Belphégor into leaving is to tell him a falsehood. He has been sent to earth for a period of ten years to investigate the institution of marriage. Honnesta, the woman he marries makes his life miserable:

> Toujours débats, toujours quelque sermon
> Plein de sottise en un degré suprême.
> Le bruit fut tel que Madame Honnesta
> Plus d'une fois les voisins éveilla.
> (ll. 129-32)

La Fontaine as author-narrator uses the tale of Belphégor as a vehicle for expressing some of his personal pessimism about marriage:

> Chez les amis tout s'excuse, tout passe;
> Chez les amants tout plaît, tout est parfait;
> Chez les époux tout ennuie et tout lasse.

Le devoir nuit: chacun est ainsi fait.
Mais, dira-t-on, n'est-il en nulles guises
D'heureux ménage? Après mûr examen,
J'appelle un bon, voire un parfait hymen,
Quand les conjoints se souffrent leurs sottises.
(ll. 117-24)

Having endured the insufferable presence of his wife's whole family and, inevitably, finding himself pursued by debtors, Belphégor abandons Honnesta and hides from her by inhabiting other bodies, a secret he confides to Matheo. Because Matheo offers him protection, Belphégor offers him the power to evict him three times from these bodies. Clever as he is, the devil is no match for Matheo. During the culminating dramatic scene of "Belphégor," Matheo, threatened with hanging if he does not exorcize Belphégor's fourth victim, devises an effective stratagem. To the accompaniment of ominous drums, he lies by announcing the arrival of Honnesta:

Il dit tout bas qu'on battît le tambour,
Ce qui fut fait; de quoi l'Esprit immonde
Un peu surpris au Manant demanda:
Pourquoi ce bruit? coquin, qu'entends-je là?
L'autre répond: C'est madame Honnesta
Qui vous réclame, et va par tout le monde
Cherchant l'Epoux que le Ciel lui donna.
Incontinent le Diable décampa,
S'enfuit au fond des Enfers...
(ll. 255-63)

For both Belphégor and the nameless devil in the land of Papefiguière,[8] a defensive lie produces their counterfactual framing of the situation as a dangerous one and leads to their decision to depart.

In "Le Cuvier"[9] Anne hides her lover under a barrel when her husband arrives home at an inopportune moment:

Ils en étaient sur un point, sur un point...
C'est dire assez de ne le dire point,
Lorsque l'époux revient tout hors d'haleine
Du cabaret, justement, justement...
(ll. 23-26)

The repetition of the word "point," even to the extent of visually representing "points" (periods) at the end of two lines, focuses attention on the sexual wordplay. With little time to reflect, Anne has to invent a defensive lie when her husband announces that he has sold the barrel for 15 francs. She quickly states that she has sold it for a higher price to a buyer who is at that very moment examining it. This strategic "lie in a crisis" protects Anne and her lover from discovery. Her husband consequently sees the situation through a false cognitive frame. His counterfactual framing prevents him from seeing the infidelity. In his eyes the lover is a buyer and his wife is an astute businesswoman. So blind is he to the real situation that he makes himself physically blind as well by going into the barrel to examine it in detail before concluding the deal. The cognitive blindness coupled with the physical blindness provides a further opportunity to dupe the husband. While the husband is occupied inside the barrel preparing it for the customer, his wife is busy outside the barrel continuing the sexual activity that had been interrupted by the unexpected return of her husband. Thus the original defensive lie uttered to avoid harm functions also to produce a benefit to the two lovers.

THE PROTECTIVE LIE

A subcategory of the defensive lie is the protective lie, a legitimate subterfuge of the kind used in circumstances where the truth is considered harmful to the well-being of, for example, a medical patient, or even a close friend. A lie of this nature is found in "Les Oies de frère Philippe," a tale whose freshness and charm recalls, as Vittorio Lugli puts it, the La Fontaine of the *Fables*:

> E la storia ingenua ha una frechezza di boschi, un'armonia di uccelli canori, da ricordare il nuovo scolaro di Esopo, il suo libro da tre anni uscito" (127).
> [And the ingenuous tale has a freshness of the woods, a harmony of singing birds, that recalls the young disciple of Aesop, his book published three years previously.—my translation]

"Les Oies de frère Philippe" is a male rendition of the *école des femmes* motif. A solicitous father raises his child away from civilization according to a graduated educational program, beginning with flowers and animals at five years old, adding at ten years of age information about the afterlife, and at fifteen introducing him to the concept of God. The one taboo subject that he methodically avoided is the existence of women. Philippe finally takes his twenty-year-old son on an excursion to the nearest village. Very curious, the son is busy identifying and categorizing all he sees:

> Le jeune homme tombé des nues
> Demandait: Qu'est-ce là? — Ce sont des gens de cour.
> — Et là? — Ce sont palais.— Ici? — Ce sont statues.
> (ll. 142-44)

He answers honestly all his son's questions, save one:

> Qu'est-ce là, dit-il à son père,
> Qui porte un si gentil habit?
> Comment l'appelle-t-on? Ce discours ne plut guère
> Au bon vieillard, qui répondit:
> C'est un oiseau qui s'appelle oie.
> (ll. 153-57)

The deception does not completely succeed because the boy's natural desires have been awakened, even though his cognitive perception has been foiled so that women are amusingly put into his already existent category of birds:[10]

> O l'agréable oiseau! dit le fils plein de joie.
> Oie, hélas chante un peu, que j'entende ta voix.
> Peut-on point un peu te connaître?
> Mon père, je vous prie et mille et mille fois,
> Menons-en une en notre bois,
> J'aurai soin de la faire paître.
> (ll. 158-63)

THE WHITE LIE

Utilitarian thinkers categorize some defensive lies as acceptable simply because their intended result is to avoid a greater harm or to achieve substantial benefit. One type of defensive white lie consistently condoned by La Fontaine's narrator is a woman's lie about her sexual activity. In "Nicaise" the author-narrator overtly upholds a woman's right to lie about her loss of virginity:

> Quelle chose? C'est celle-là
> Que fille dit toujours qu'elle a.
> Je le crois; mais d'en mettre jà

> Mon doigt au feu, ma foi je n'ose:
> Ce que je sais, c'est qu'en tel cas
> Fille qui ment ne pèche pas.
> (ll. 223-28)

In "Les Aveux indiscrets" his advice to temporarily wayward husbands and wives is to not tell their spouses: "Mentir alors est digne de pardon" (l. 115). This is blameless mendacity. In "La Fiancée du roi de Garbe," Alaciel, after all her misadventures, "conta tout ce qu'elle voulut. / Dit les mensonges qu'il lui plut" (ll. 773-74).

SILENCE

Speaking the truth when one should remain silent is a transgression of the code of love, a violation of the secrecy called for in certain circumstances. As a method of defense, silence may be used to forestall harmful effects by generating counterfactual protective belief spaces, just as a lie does.

The destructive results of revealing the truth are manifest in "Les Aveux indiscrets," a tale with no identifiable source. La Fontaine presents only the explosive dénouement to the tranquil period of these marital lives that have been functioning well under the deceptive cover of silence. As the tale opens both the husband and wife break their silence to reveal secrets from the past. First, Damon, in a burst of honesty, confesses to having fathered a daughter before their marriage:

> J'en dis ma coulpe, et j'en suis tout honteux.
> Il m'en souvient, la nymphe était gentille,
> Au fond d'un bois, l'Amour seul avec nous;
> Il fit si bien, si mal me direz-vous,
> Que de ce fait il me reste une fille.
> (ll. 26-30)

Aminte, in turn, confides:

> J'étais un jour seulette à la maison;
> Il me vint voir certain fils de famille,
> Bien fait et beau, d'agréable façon;
> J'en eus pitié; mon naturel est bon:

Et, pour conter tout de fil en aiguille,
Il m'est resté de ce fait un garçon.
(ll. 32-37)

The immediate result in this patriarchal society is the husband's changed cognitive perspective of himself—not of his wife. He has become disgraced: "Je suis bâté," he cries. Aminte's mother does not intentionally break her protective silence about her own past, but in counselling her daughter, she is overheard by her husband. He reacts in a manner identical to Damon. The discourse of Aminte's mother is, with the voice of the narrator, the expression of reason and a guide to the *bienséances*. She explains to Aminte that she has transgressed the code of secrecy required in a marital relationship:

Vous avez tort: j'ai vu beaucoup de sots,
Et plus encor de sottes, en ma vie;
Mais qu'on pût voir telle indiscretion,
Qui l'aurait cru? Car enfin, je vous prie,
Qui vous forçait? quelle obligation
De révéler une chose semblable?
(ll. 59-64)

If the defensive lie creates a counterfactual belief space and results in avoidance of harm, on the contrary, telling the truth in this situation generates a factual frame of understanding that leads to harm. The juxtaposed cognitive frames of "Les Aveux indiscrets" generate each other and prompt the narrator to comment by way of cognitive closure:

Mon discours donc en deux points se renferme.
Le noeud d'hymen doit être respecté,
Veut de la foi, veut de l'honnêteté:
Si par malheur quelque atteinte un peu forte
Le fait clocher d'un ou d'autre côté,
Comportez-vous de manière et de sorte
Que ce secret ne soit pas éventé.
(ll. 107-13)

Defensive lies and silence play a role in the cognitive closure of a number of tales, much as offensive lies prepare for the primary deceit at the beginning of some narratives. Silence protects the wife of Teudelingue from knowing she has been seduced by the muledriver ("Le Muletier"). Alaciel in "La Fiancée du roi de

Garbe" maintains a calculated silence at the end of her adventures. At the conclusion of "Le Psautier," the sexual transgressions of the nuns are cloaked in a protective silence:

> Le lendemain ne fut tenu, pour cause,
> Aucun chapitre; et le jour ensuivant
> Tout aussi peu. Les sages du couvent
> Furent d'avis que l'on se devait taire;
> Car trop d'éclat eût pu nuire au troupeau.
> (ll. 134-38)

***

Lies, of one kind or another, have been part of the lover's arsenal of seductive rhetoric from ancient times. As generators of counterfactual belief spaces, the lies in La Fontaine's tales participate in the fabrication of that ludic space in which reader and narrator meet on the cognitive level of the text to enjoy the "bons tours des maris et des femmes" ("Joconde" l. 499). Like the jocose lie of St. Thomas, the *Contes* fabricate a universe that is ludic in intention and meant to please, not to harm: "Qui ne voit que ceci est jeu, et par conséquent ne peut porter coup?" (Preface to the *Contes*).

NOTES

1. See Sissela Bok's definition in *Lying: Moral Choice in Public and Private Life* : "[A lie is] any intentionally deceptive message which is *stated.* Such statements are most often made verbally or in writing, but can of course also be conveyed via smoke signals, Morse code, sign language, and the like. Deception, then, is the larger category, and lying forms part of it" (13-14).

2. The two most influential theologians on the subject of lying have been Augustine and Thomas Aquinas. See Augustine's *Enchiridion* as well as "On Lying" and "Against Lying" in his *Treatises on various subjects* (vols. 14 and 15) where he sets up an eightfold categorical system and distinguishes lies on the basis of intentionality and the harmfulness of their effects. Thomas Aquinas, *Summa Theologica* (2.2 ques. 109,110) differentiates officious or helpful lies from jocose lies, told in jest, and from malicious lies told to harm someone. See Bok's chapter 3: "Never to Lie?"

3 "On ne s'avise jamais de tout" is likewise based on a misleading signal. The wife of a pathologically jealous husband deceives him when her lover throws garbage on her as she passes under his window on the way home from church. As her female companion explains to the husband, she had to go into the house to change her clothes. The husband immediately recognizes the lie—"je suis pris pour dupe."

4 I am indebted to Scheibe, *Mirrors, Masks, Lies and Secrets* for the term "intrusive lie" (82).

5 Boccaccio's version (*Decameron* III, 3) is much more cumbersome and moralizing, perhaps because the *entremetteur* is a priest.

6. While one has the impression that flattery plays a significant role in the *Fables*, in reality flattery is a major deceptive device in only a few fables—"Le Corbeau et le Renard" (I, 2), "Les Animaux malades de la peste" (VII, 1), "Les Obsèques de la lionne" (VII, 14), "Le Renard, le Loup et le Cheval" (XII, 17).

7. The source is Boccaccio, *Decameron*, III,1. At the end of Boccaccio's tale the abbess discovers the seducer can speak; then the nuns decide to keep the episode a secret by declaring his "cure" a miracle due to their prayers. In La Fontaine the nuns seem to believe in the miracle, but decide to keep him in order to preserve their secret.

8. The trickery of the devil with an enigmatic pubic hair in "La Chose impossible" could be considered the equivalent to defensive mendacity in that the dupery plays on the devil's lack of knowledge and aims to protect the deceivers.

9. The source is Boccaccio, *Decameron* (VII, 2).

10. John D. Lyons finds in this tale a significance going well beyond the immediate perception of an amusing narrative: "Chez La Fontaine comme chez Boccace cette nouvelle concerne le regard masculin sur les femmes et les difficultés de faire entrer la réalité de celles-ci dans le discours des hommes. Dans l'histoire du frère Philippe il n'est pas question de parler *aux* femmes mais de parler *des* femmes ou de la femme en général. La femme-oie, c'est l'absence de mot dans le langage masculin pour représenter le monde où il y a des femmes" (118).

CHAPTER SIX

FALSE PROMISES: A NARCISSISTIC TRAP

Seductive promises are subversive performative speech acts that actually are, even on the linguistic level, traps because of their ultimate self-referentiality:

> The trap of seduction thus consists in producing a *referential illusion* through an utterance that is by its very nature *self-referential*: the illusion of a real or extralinguistic act of commitment created by an utterance that refers only to itself. (Felman 31)

A promise exploits the capacity of language to reflect upon itself. Promises fit into the linguistic category of performatives. In other words, they accomplish at the time of their utterance what they say they are doing. A "false" promise is really an insincere promise, what J. L. Austin terms an infelicitous promise, one that the promiser knows will not be carried out (239). Close to lies in that they maintain secrecy about the true intentions of the seducer, false promises wield their true performative power by setting up false belief spaces about the future in the mind of the victim. Such acts of promising are hollow and purely self-referential. All they ultimately say on the linguistic level is: "I am promising." Many of La Fontaine's *Contes*, like their antecedents, the *fabliaux* and the tales from the Italian and French traditions, are plainly the elaborations of male sexual fantasies about the stupid—or "innocent"—young woman who is easily seduced by the verbal manipulation of a clever man. The act of manipulation as it is represented in the narrative aims at falsifying the victim's way of perceiving her present situation, her future goals, and the means of attaining them. Basing her decision on the false belief spaces that have been engendered, the victim falls into the trap and the seducer achieves his goal. The belief spaces set up by the false promise constitute a misrepresentation of the future, in what I term the hypothetical mode, because the promise is grounded on a hypothesis: *if* you sleep with me, you will become intelligent. It is a false exchange in which an unverifiable future result is exchanged for a known present act.The pattern of deceit can be considered as an embedded anticipation frame; the deceivee's own hope for the fulfillment of the promise is embedded within a false framing of the situation.

Usually several identifiable stages occur in the structuring of the cognitive pattern of a tale based on manipulation by means of false promises. In the introductory phase, the seducer's goal, or the woman's quest, is established, while at the same time the naiveté of the victim is underlined.[1] A preparatory lie frequently produces the first of the deceivee's false belief spaces—a falsely perceived present situation. By this stage a one-sided fiduciary relationship exists without which the next phase cannot succeed. The next step is the seducer's false promise of a means by which to change the situation that now is viewed as undesirable. The false promise produces the victim's construction of a hypothetical future goal and the means to attain it. She believes the promised result will ensue from the sexual act. She, like the seducer, *hopes* for a benefit. Next, the deceivee fulfills her part of the exchange by acceding to the promiser's sexual desires. The deconstructive phase follows quickly upon the victim's decision and her involvement in the sex act. The seducer has achieved his goal, but the deceivee does not receive the expected benefit, that is, the fulfillment of the promise. With the discovery of the insincerity of the promise— although it does not occur in all cases—comes the final stage and the cognitive closure: the total collapse of the previously constructed belief spaces.

In "Le Faiseur d'oreilles et le raccommodeur de moules"[2] the narrative development can be seen as a series of acts progressively producing false belief spaces, that in turn occasion further acts. The overall structure is a deceit pattern juxtaposing two complementary false promises and two sexual transgressions that function on the model of a patriarchal type of exchange (your wife for mine). The first 17 lines establish the belief context, or fiduciary relationship, within which the preparatory lie will be believed and the false promise will be accepted, and then acted upon. Sire Guillaume has gone on a business trip leaving at home his young wife Alix who is six months pregnant and the epitome of the ignorant victim. She is "fort neuve" (l. 12), unencumbered by too much intelligence, states the narrator. André, the seducer, on the other hand, is an expert at seduction: "Il ne tendait guère en vain ses filets" (l. 8). A preparatory lie leads Alix to believe that her unborn child is missing an ear:

Je m'ébahis comme au bout du royaume
S'en est allé le compère Guillaume,
Sans achever l'enfant que vous portez:
Car je vois bien qu'il lui manque une oreille.
(ll. 22-25)

The counterfactual belief space is now in place. Upon this lie the deceitful promise is built, the promise that through sexual intercourse André will supply the missing ear:

. . . Allez, n'ayez souci,
Répliqua-t-il; je prends sur moi ceci.
(ll. 40-41)

André's promise to "complete" the formation of Alix's unborn child is, as Benveniste would state, "infelicitous." It succeeds as rhetorical performance and it succeeds as a subversive speech act leading to moral/social transgression. It does not truly succeed, however. It is "infelicitous" because the promise of adding a missing ear to the baby is insincere and the seducer knows it cannot be fulfilled. It has no extralinguistic point of reference. The promise engenders in Alix's mind a false belief space in that she now perceives a desirable future goal, hypothetical in nature: the restoration of her child's ear. Believing that the means proposed will result in the attainment of her goal, she therefore cooperates wholeheartedly in the sexual "game" (ll. 43-71). André carries out meticulously the process of creating the missing ear:

André vaquait de grande affection
A son travail; faisant ore un tendon,
Ore un repli, puis quelque cartilage.
(ll. 46-48)

In fact, André pursues his work with such devotion that by the next day Alix becomes concerned about whether her child may not end up with too many ears! When Guillaume returns, he discovers the deceit from his wife who reproaches him for having left without finishing their child's ear.

The second half of the *conte* demonstrates a false promise functioning as a "seductive" tool in quite a different manner, for Guillaume's goal is really revenge, not seduction. His first thought is to cut off one of André's ears, or worse: "Peut-être pis, ce qu'on coupe en Turquie" (l. 131). Then, a better idea comes to mind. In the presence of the concealed husband, Guillaume promises

André's wife that he will "correct" the defective "mould" (in twentieth-century terms, the defective gene) that gives all her children too short a nose. On the surface, this false promise resembles the earlier one made to Alix, with its preparatory lie, followed by the promise of a future benefit. André's wife, however, is not naive; she does not believe in the defective mould; she does not expect the promise to be fulfilled. Furthermore, Guillaume does not expect her to believe his claim. The false belief space of the first deception, that of Alix, is simple, while that of the second deception is more complex. In the second situation the deceiver, Sire Guillaume, pretends not to know that André is present and pretends that out of gratitude for André's service, he wants to offer a similar benefit in return. The irony of this promise is meant to be comprehended by both André and his wife:

> L'ingratitude est mère de tout vice.
> André m'a fait un notable service;
> Parquoi, devant que vous sortiez d'ici,
> Je lui rendrai, si je puis, la pareille.
> (ll. 166-69)

The promise is more "felicitous" in this case because the implicit promise, made to both André and his wife, is that her consent to the sexual demands of Guillaume will constitute complete revenge for the outrage to Guillaume's honor. The threat to André's ears—and other appendages—will be cancelled. This is in fact exactly what takes place:

> André vit tout, et n'osa murmurer;
> Jugea des coups; mais ce fut sans rien dire;
> Et loua Dieu que le mal n'était pire.
> Pour une oreille il aurait composé.
> Sortir à moins c'était pour lui merveilles:
> Je dis à moins; car mieux vaut, tout prisé,
> Cornes gagner que perdre ses oreilles.
> (ll. 195-201)

In the diagram below, André's framing of the manipulation in the first part (ll. 20-81) is factual; Alix's false hope—framed in exclamation marks—is embedded in her counterfactual interpretation of the situation. André knows that his lie produces a false and counterfactual belief space for Alix; he also understands that the false promise creates her framing of a hypothetical desirable state in which her child will have two ears.

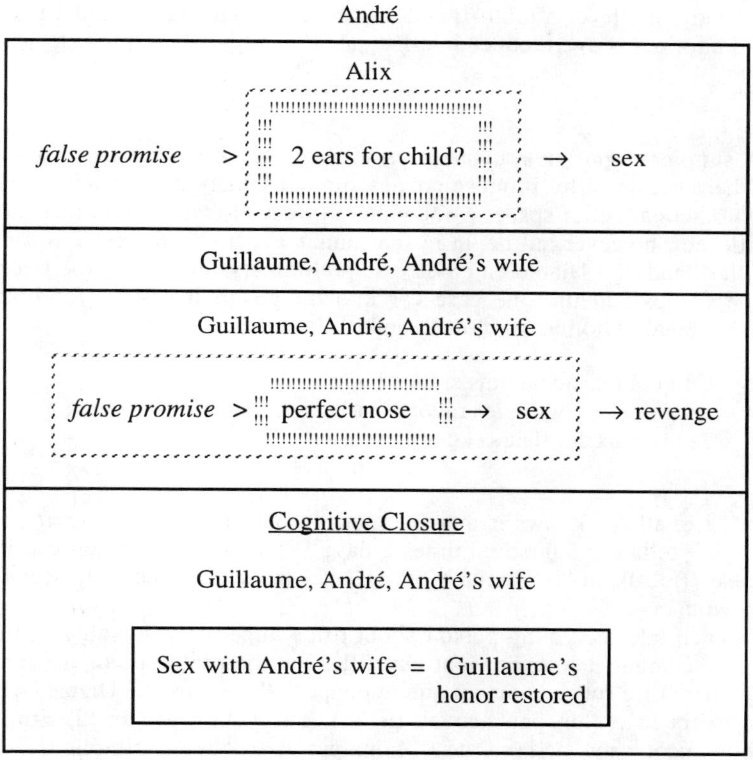

Figure 6.1

In the second part (ll. 82-201) Guillaume, André, and André's wife all understand clearly the nature of each other's belief spaces. Guillaume's wife, Alix, stands by uncomprehending and is not a part of the cognitive situation. The other three juggle in their cognitive space two framings of the same action. They comprehend that they are functioning within an assumed false belief space; this understanding acts as a filter of pretense through which the action is interpreted. Each one pretends to believe—and knows that the others are pretending to believe—that an act of seductive manipulation is taking place. The cognitive closure represents what they really see. They know that this is not a scene of seduction, a manipulation through a false promise. It is, rather, a simple exchange in which each party understands clearly that what is going on is a primitive "eye for an eye and a tooth for a tooth" pattern of revenge. This latter model is covert, while the former is overt, but counterfactual.

***

A simpler cognitive structure presents itself in three tales of sexual initiation where the deceitful promise creates simultaneously the counterfactual and the hypothetical belief spaces.[3] The pattern of deceit is identical in each *conte*. They do not, however, all begin in the same way. In "Comment l'esprit vient aux filles" and "Le Diable en enfer" the preliminary impetus of the tale is the heroine's quest, in the one case for knowledge, in the other for holiness. Alibech's ideal of holiness is not altogether altruistic:

> O! quel plaisir j'aurais si tous les ans,
> La palme en main, les rayons sur la tête,
> Je recevais des fleurs et des présents!
> (ll. 46-48)

As for Lise, all she knows is that she must go out to buy some *esprit* because her mother tells her a hundred times a day: "Va-t'en chercher de l'esprit malheureuse" (l. 30). In "L"Ermite" (pp. 687-92) the seducer actively seeks out a young woman.

In each tale the young person about to be duped is vulnerable and naive. Lise of "Comment l'esprit vient aux filles" is the prototype of the innocent young girl: "Lise songeait autant que sa poupée" (l. 28). In "Le Diable en enfer" young Alibech is "un peu neuve" (l. 26). The young girl in "L'Ermite" is "Pucelle encore; mais à la vérité / Moins par vertu que par simplicité" (ll. 28-29). This then is the belief context that makes the false promise so immediately credible.

The deceit in the first two tales takes place when the woman meets a seducer who assures her that the sexual act is the means to achieve her goal. The seducer, who is in her eyes a highly respected religious person, presents himself as the possessor of arcane knowledge. In "Le Diable en enfer" the hermit, Rustic, unlike Bonaventure in "Comment l'esprit vient aux filles," is not hypocritical or conniving at the outset. He does not indulge in a preparatory lie. His downfall is self-deception. He thinks he is strong enough to resist temptation and sees her staying overnight with him as a challenge to increase his holiness. His interior monologue betrays his foolhardiness:

Il dit en soi: Rustic, que sais-tu faire?
Veiller, prier, jeûner, porter la haire?
Qu'est-ce cela? moins que rien, tous le font:
Mais d'être seul auprès de quelque belle
Sans la toucher, il n'est victoire telle.
(ll. 87-91)

Rustic resorts to deceit only when, seeing Alibech sleeping in the middle of the night, the temptation becomes too great. He invents a falsehood that is fundamentally a promise: by means of the sex act (an activity she knows nothing about) she will become holy, she will overcome the devil, "Emprisonnant en enfer le Malin" (l. 134). Alibech's consent is based on a false cognitive frame: "Elle crut faire acte de repentante" (l. 144). Bonaventure in "Comment l'esprit vient aux filles" is, on the other hand, deceitful from the outset. Upon hearing why Lise has come to see him—she seeks *esprit*—he assures her in a preparatory lie that he can indeed sell her what she desires:

Nous pourvoirons à ce qui vous amène,
Sans exiger nul salaire de vous:
Il est marchand et marchande, entre nous:
A l'une on vend ce qu'à l'autre on donne.
(ll. 58-61)

While "Comment l'esprit vient aux filles" would seem to be a more original creation by La Fontaine because a specific source has not been found, the poetic style of "Le Diable en enfer" is more inventive, and owes little to its source, Boccaccio's *Decameron* (II, 10). A comparison of the manner in which the two texts of La Fontaine suggest the sexual act without being too explicit shows that in "Comment l'esprit vient aux filles" the metaphor of "l'esprit" is the chief evocative vehicle, while in the other tale, even the rhythms of the verse are highly mimetic. In the first text there is no textual foreplay; the moment is brief:

Il suit sa pointe; et d'encor en encor
Toujours l'esprit s'insinue et s'avance,
Tant et si bien qu'il arrive à bon port.
(ll. 76-78)

In "Le Diable en enfer," however, La Fontaine summons up his resources of metaphor, alliteration, anaphoric repetition, and rhythmic patterns:

> . . . Une certaine bête
> Diable nommée, un vrai serpent maudit,
> N'eut point de paix qu'il ne fût de la fête.
> On l'y reçoit. Rustic roule en sa tête,
> Tantôt les traits de la jeune beauté,
> Tantôt sa grâce, et sa naïveté,
> Et ses façons, et sa manière douce,
> L'âge, la taille, et surtout l'embonpoint,
> Et certain sein ne se reposant point;
> Allant, venant; sein qui pousse et repousse
> Certain corset en dépit d'Alibech,
> Qui tâche en vain de lui clore le bec:
> Car toujours parle: il va, vient, et respire.
> C'est son patois; Dieu sait ce qu'il veut dire.
> (ll. 112-25)

The rhythm, alternatively halting and sinuously penetrating, suggests the intermingling movements of a penis, an obsessive temptation, and the object of seduction represented by a woman's breast. When the moment of poetic foreplay dissipates, the text, like that of "Comment l'esprit vient aux filles," relies highly on its central metaphor to convey the sexual act, albeit in a much lengthier and detailed passage (ll. 137-69).

The cognitive closure of "Comment l'esprit vient aux filles" differs from the ending of "Le Diable en enfer" in that the promise is, in a way, fulfilled in the first tale. Alibech did not find saintliness, but Lise does find *esprit*, a new cognitive state, through her sexual initiation. She suddenly begins to think, startling even the implied reader into disrupting the tale:

> Lise lui fait une humble révérence;
> Et s'en retourne en songeant à cela.
> Lise songer! Quoi? déjà Lise songe!
> Elle fait plus: elle cherche un mensonge,
> Se doutant bien qu'on lui demanderait,
> Sans y manquer, d'où ce retard venait.
> (ll. 96-101)

This is the traditional *école des femmes* motif, as the narrator comments in the introduction to the tale:

Avant que Lise allât en cette école,
Lise n'était qu'un misérable oison.
Coudre et filer c'était son exercice;
Non pas le sien, mais celui de ses doigts;
Car que l'esprit eût part à cet office,
Ne le croyez; il n'était nuls emplois
Où Lise pût avoir l'âme occupée:
Lise songeait autant que sa poupée.
(ll. 21-28)

Alibech in "Le Diable en enfer" follows a similar cognitive evolution for, as her friends point out, she could have learned the same lesson without going off to find a hermit: "On vous aurait, sans bouger du logis, / Même leçon, même secret appris" (ll. 192-93).

In "L'Ermite" the meeting with the seducer is wholly initiated by him. Like the hermit in "Comment l'esprit vient aux filles," the seducer, Luce, is a masterful hypocrite. He has a carefully honed image of holiness and chastity. He hides when he sees a woman: "Se renfermait voyant une femelle," practicing the monastic virtue of custody of the eyes: "et baissait la prunelle" (ll. 20-21). Luce fixes his desires on a young girl who lives with her widowed mother, and he subsequently elaborates an ingenious deception scheme. The preliminary deceit that establishes an initial framework of credibility is a disguise that continues to falsify the perspective throughout the tale; a voice issuing from a purportedly divine source in the middle of the night decrees in prophetic terms:

Réveillez-vous créatures de Dieu,
Toi femme veuve, et toi fille pucelle:
Allez trouver mon serviteur fidèle
L'ermite Luce, et partez de ce lieu
Demain matin sans le dire à personne;
Car c'est ainsi que le Ciel vous l'ordonne.
(ll. 57-62)

In a parody of the biblical promise of the angel Gabriel to Mary, the voice promises that if the daughter goes to see the hermit, she will conceive a wondrous child:

Car d'eux doit naître un pape, dont la vie
Réformera tout le peuple chrétien.
(ll. 66-67)

The structure of the deception is different from other tales involving false promises made to naive women in that the belief in the false promise is progressively built up in "L'Ermite." The resistance to accepting the veracity of the promise centers first on correctly identifying the emitter of the oracle. La Fontaine deviates from his source, *Les Cent nouvelles nouvelles*, in having the daughter, as well as the mother, hear the mysterious oracle. This permits further emphasis on the innocent vulnerability of the girl who, unlike her mother, is more willing to obey the directive immediately. The mother wonders whether the promise might not be from the devil, or whether their fear might have caused a misunderstanding of the words (ll. 87-98). The fiduciary contract with the seducer is mediated by the mother and develops in a tripartite movement. The first enunciation of the promise is followed by disbelief, but ends with the mother and daughter praying to ward off the possibility of trickery by the devil. In the next stage the oracle, in a second pronouncement, threatens death for disobedience to the instructions:

> Femme incrédule et qui vas à l'encontre
> Des volontés de Dieu ton créateur,
> Ne tarde plus, va-t'en trouver l'ermite,
> Ou tu mourras. . . .
> (ll. 107-10)

This time they act, but still with circumspection. Having found the hermit, they seek corroboration of his trustworthiness by peeking through a hole he has conveniently made in the wall of his dwelling. The cognitive framing is illusory. The scene framed through this peephole shows a saintly man practicing self-flagellation, a sure sign of holiness and fervor in those times. The scene, however, has been cleverly staged by the hypocritical monk who is actually not beating himself at all. He makes it appear and sound as if he is scourging himself with fifty disciplines:

> Et par un trou qu'il avait fait exprès
> A sa cellule, il voulait que ces femmes
> Le pussent voir comme un brave soldat
> Le fouet en main, toujours en un état
> De pénitence, et de tirer des flammes
> Quelque défunt puni pour ses méfaits,
> Faisant si bien en frappant tout auprès
> Qu'on crût ouïr cinquante disciplines.
> (ll. 121-28)

His imposture continues as he reluctantly opens the door and appears astonished at the news they bring to him:

> Je crains, dit-il, les ruses du malin:
> Dispensez-moi, le sexe féminin
> Ne doit avoir en ma cellule entrée.
> (ll. 140-42)

A magical third time the voice of the oracle resounds that night, assuring them that Luce has had a change of heart. Mother and daughter return to the hermit the next day. The daughter remains with him for about seven months, becoming more intelligent each day as she learns from her lovemaking, from the "arbre de science" (l. 176), as the narrator calls it.

The tale concludes with the anticlimactic birth of a daughter. The discovery of the false nature of the promise, through its infelicitous results brings about a cognitive closure where, just as in the fable of "La Laitière et le pot au lait," the whole futuristic dream scaffold collapses. Both mother and daughter are left with a retroactively true interpretation of the preceding events:

> Mais ce qui vint détruisit les châteaux,
> Fit avorter les mitres, les chapeaux,
> Et les grandeurs de toute la famille:
> La signora mit au monde une fille.
> (ll. 205-08)

All three of the preceding tales show a deceivee falsely perceiving a deficiency, while the false promise engenders a false hope in a hypothetical, somewhat utopian, future situation in the mind of the deceivee:

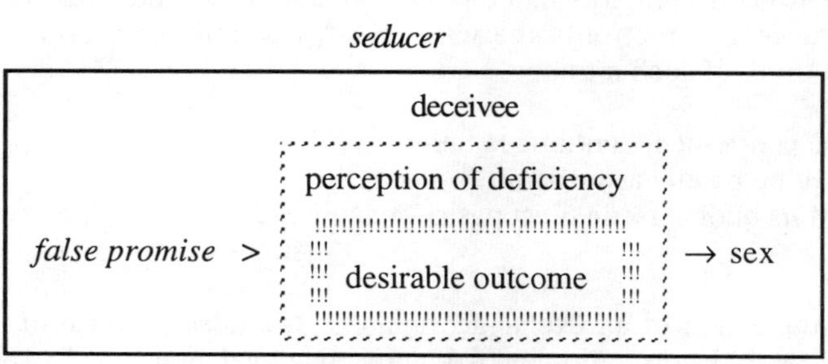

Figure 6.2

***

A false promise has a role to play, as well, in three *contes* where exchanging goods or money for sexual favors is the basic narrative program; one side of the bargain turns out to be a false promise that subverts the planned exchange.

In "Conte d'une chose arrivée à Château-Thierry," a tale thought to originate from an event in La Fontaine's own town of birth, the action begins with a creditor who approaches a woman with a proposal to forgive a debt that she and her husband, Blaise, owe him, if she yields to his desires. In a conspiratorial tone evoked textually by the parenthesis, he suggests:

> Vous avez trop de quoi me satisfaire,
> (Ce lui dit-il) et sans débourser rien:
> Accordez-moi ce que vous savez bien.
> (ll. 10-12)

The proposed exchange is between these two; the agreement is, however, insincerely entered upon by the wife, who swiftly informs her husband. The husband and wife then enter into collusion to dupe the creditor. The main scene resembles the scene in Molière's *Tartuffe* where the husband, Orgon, hides under the table waiting for his wife, Elmire to cough and alert him before Tartuffe's seduction goes too far. In this tale, the woman takes over the initiative and falsely promises to perform her part of the exchange if the creditor first relinquishes the credit note. Blaise and his wife understand the total cognitive situation, knowing that because of the wife's false promise the creditor misinterprets the exchange pact and foresees in his hypothetical belief frame the hope of a sexual encounter. In a coda to the brief tale La Fontaine adds a social commentary, much in the style of the moral at the end of a fable like "Les Animaux malades de la peste." A "gros bourgeois" remarks that the wife should have waited to cough until *after*, but she replies that only the wife of a "gros bourgeois" would be clever enough to think of such a thing:

> Je pense bien, continua la belle,
> Qu'en pareil cas madame en use ainsi;
> Mais quoi, chacun n'est pas si sage qu'elle.
> (ll. 39-41)

The subversion of an exchange, again by the false promise of money for sex, is central to the deception in "A Femme avare galant escroc."[4] The exchange is to take place between Gulphar and a wife so avaricious that she is willing even to be unfaithful to her husband for the right sum of money. If her goal is money, his is sex. I am inclined not to see this tale as presenting a *trompeur trompé* pattern of deceit, except as a secondary motif, because her objective is not sexual; she strikes a simple bargain with Gulphar and thus provides him with the opportunity to deceive her. This time the exchange is subverted not by the wife, but by Gulphar, her would-be lover. He maneuvers her into a situation where she would have to confess her infidelity to her husband in order to keep the money that was to be her recompense. The wife is duped when Gulphar obtains from her husband, Gasparin, a loan of 200 *écus*, the exact sum his wife had just accepted before "paying" Gulphar generously in her own way:

> Le lendemain elle le régala
> Tout de son mieux, en femme de parole.
> Le drôle en prit ce jour et les suivants
> Pour son argent, et même avec usure:
> A bon payeur on fait bonne mesure.
> (ll. 50-54)

Gulphar had given the money to Madame in front of others, feigning, so she thought, that she was to return it to her husband. Her bluff is called, however, when Gulphar tells Gasparin, with his wife present, that she has the money he borrowed:

> J'ai votre argent à Madame rendu,
> N'en ayant eu pour une affaire urgente
> Aucun besoin, comme je l'avais cru.
> (ll. 57-59)

The husband, as secondary deceivee, never does discover his wife's avaricious scheme. While the creditor in "Conte d'une chose arrivée à Château-Thierry" did not get the sexual payment and also lost his credit note, here, the wife does not get her money and she has also paid out her part of the bargain. The extension of the cognitive frame in the final lines, so that the whole village shares the same knowledge constitutes a recognition of the tale's grounding in the reality of French society:

> En la quittant, Gulphar alla tout droit
> Conter ce cas, le corner par la ville,
> Le publier, le prêcher sur les toits.

De l'en blâmer, il serait inutile:
Ainsi vit-on chez nous autres François.
(ll. 67-71)

"Le Magnifique" also displays the pattern of embedded anticipation created by a false promise, but this tale incorporates as well the juxtaposed frames of a *trompeur trompé* deceit pattern. Aldobrandin, an overly jealous husband, promises in exchange for le Magnifique's horse that the man can spend a quarter of an hour with his wife—but within Aldobrandin's sight. The would-be lover soon discovers that the promised exchange has been sabotaged by Aldobrandin who has forbidden his wife to speak during the interview:

Je vois, je vois, c'est une tricherie
De votre époux: il m'a joué ce trait;
Et ne prétend qu'aucune repartie
Soit du marché: mais j'y sais un secret.
(ll. 124-27)

Le Magnifique, nevertheless, manages to thwart the devious husband's scheme by inventing a long dialogue with the woman in which he, in a veritable dramatic *tour de force*, speaks for both of them, persuading her to meet him in the garden later that night. As they part, through his voice he puts on her lips an avowal of love:

. . . Ah mon cher Magnifique
Que je vous aime! et que je vous sais gré
De ce dessein! venez, je descendrai.
(ll. 159-61)

The lover subverts the husband's scheme by conjuring up a hypothetical anticipatory space within which he imagines the woman's reactions. Just as in "A Femme avare galant escroc," the tale concludes with the husband still completely unaware that he has been duped.

***

Finally, "L'Oraison de saint Julien" presents a false wager being undermined so that the hero, Renaud, finds the answer to his prayer in the arms of a beautiful widow. When Renaud meets three fellow travellers, one of whom is highly skeptical about the power of Renaud's prayer to St. Julian for a good lodging, he obligingly accepts a wager as to which one of them will find a better bed that night:

> Gageons un peu quel sera le meilleur,
> Pour ce jour d'hui, de mon gîte ou du vôtre.
> (ll. 64-65)

Should the new companion lose, he will begin to pray to the saints; should Renaud lose, he will give up his horse, his clothing, and his money. In fact, the wager was proposed insincerely. He has fallen among thieves who immediately rob him and leave him in a pitiful state far from any place to pass the night:

> . . . et le pauvre Renaud,
> En caleçons, en chausses, en chemises,
> Mouillé, fangeux, ayant au nez la bise,
> Va tout dolent; et craint avec raison
> Qu'il n'ait, ce coup, malgré son oraison
> Très mauvais gîte . . . .
> (ll. 97-102)

Because of the trickery Renaud arrives too late at his hotel and finds himself taken into the home of a generous woman who sees in him a fortunate replacement for the man she had intended to entertain that night:

> Homme pour homme, et péché pour péché,
> Autant me vaut celui-ci que cet autre.
> Renaud n'était si neuf qu'il ne vît bien
> Que l'oraison de monsieur saint Julien
> Ferait effet, et qu'il aurait bon gîte.
> (ll. 235-39)

La Fontaine luxuriates in the description of the bath, the perfume, the meal, the bed, all of which the source-tale, Boccaccio's *Decameron* (II, 2),[5] curtly alludes to:

> Le souper prêt, la chambre bien parée;
> Bons restaurants, champignons, et ragoûts;

Bains, et parfums; matelas blancs et mous;
Vins du coucher . . . .
(ll. 167-70)

The lovemaking scene is also much more prolonged and detailed (ll. 241-315) than in Boccaccio. An itinerary of love reminiscent of the *Carte du Tendre*—"car nul point n'y manquait" (l. 293)—takes them through *louange* (ll. 263-88), *résistance* (ll. 294- 96), *baisers* (l. 299), *la petite oie* (l. 300):

Voilà, disait la veuve charitable,
Pour le chemin, voici pour les brigands,
Puis pour la peur, puis pour le mauvais temps:
Tant que le tout pièce à pièce s'efface.
Qui ne voudrait se racquitter ainsi?
Conclusion, que Renaud sur la place
Obtint le don d'amoureuse merci.
Les doux propos recommencent ensuite,
Puis les baisers, et puis la noix confite.
(ll. 306-14)

And then, on to the *terres inconnues* of the bed, where the narrator confesses not to know what happened:

Je n'ai pas su ce qu'étant dans le lit
Ils avaient fait; mais comme avec l'habit
On met à part certain reste de honte,
Apparemment le meilleur de ce conte
Entre deux draps pour Renaud se passa.
(ll. 319-23)

The major cause of Renaud's good fortune is not so much the encounter with the insincere thieves and their false promise, as it is attributable to the success of his prayer to St. Julian. The scandal is not the broken wager but the kept promise, the contract made between Renaud and the saint, that results, somewhat sacrilegiously, in the saint's delivering to Renaud a particularly accommodating bed for the night—"Et grâce à Dieu et monsieur saint Julien / Eut une nuit qui ne lui coûta rien" (ll. 366-67).

While this tale is not like those based on deception by a false promise that leads to the seduction of a naive victim, it does share with them in its cognitive structure the reliance on discrepant states of awareness or frames of expectation to create an ironic perspective for the reader. The narrator points out in the concluding lines the double irony of the situation—the thieves' circumstances appear to be fortunate, but in reality lead to their death, while Renaud, whose plight seemed desperate, by never renouncing his cognitive frame of anticipation ends up happy:

> . . . trois gens par-devers eux
> Ont un roussin, et nombre de pistoles:
> Qui n'aurait cru ces gens-là fort chanceux?
> Aussi font-ils florès et caprioles,
> (Mauvais présage) et tout gais et joyeux
> Sont sur le point de partir leur chevance,
> Lorsqu'on les vient prier d'une autre danse.
> En contréchange un pauvre malheureux
> S'en va périr selon toute apparence,
> Quand sous la main lui tombe une beauté.
> (ll. 354-63)

***

The narcissism of the false promise consists in its autoreferential quality; its only performative power is in its own utterance and in setting up false belief spaces about an unattainable future in the mind of the victim. Unlike lies, the false promise pretends to be based on a performative contract. In the autoerotic context of seduction the use of self-referential insincere promises is particularly congruous. As Shoshana Felman so aptly puts it: "The scandal of seduction seems to be fundamentally tied to the scandal of the broken promise" (11).

NOTES

1. While I refer to sexual deception as seduction, it is often doubtful, particularly in the tales based on verbal dupery, whether an action founded on the exploitation of extremely naive victims constitutes, properly speaking, seduction. Just as Freud is said to have avoided the reality of incest and sexual abuse by terming such actions seduction, so in La Fontaine what is presented as seduction is sometimes closer to rape. Naive young women are lured into sexual encounters by falsehoods that they do not have the ability to reject: they are told that sexual intercourse is the means to obtain *esprit* ("Comment l'esprit vient aux filles"), or a tithe that must be paid to the Church ("Les Frères de Catalogne"),

or a way to complete the formation of an unborn child's ears ("Le Faiseur d'oreilles et le Raccommodeur de moules"), or how to overcome the devil ("Le Diable en enfer").

2. In Boccaccio, VIII, 8 there are no missing ears and no deficient moulds. The husband tells the woman neighbor directly that her husband is sleeping with his wife and that he is going to punish her husband by sleeping with her. She agrees and the deed is done on a *coffre* where Zeppa, the husband, is hiding. Afterwards Zeppa comes out and proposes that they now have wives in common! For the first part La Fontaine draws some inspiration fromTale 9 of Bonaventure des Périers, *Nouvelles récréations et joyeux devis* (*Conteurs français* 326-30), where André's wife, while he is away, naively gets ears put on her unborn child. When André upon his return discovers the treachery, he angrily confronts his neighbor, but is easily appeased by a gift.

3. "La Jument du compère Pierre" could also fit into this category, but I have chosen to consider it more as a tale of pseudo-magic in chapter 1. "Nicaise" presents a naive man to whom a woman makes a sincere promise—to make love to him before anyone else. But when her marriage day comes and she goes to him in the garden, he shows himself to be a true *sot* by leaving her to get a carpet so that her clothes will not get wet. The promise is well-intentioned and sincere, but a foolish one; the promise ends up revealing his stupidity and her own lack of judgment.

4. See Boccaccio, VIII, 1; Chaucer's version of the tale is "The Shipman's Tale" (*Canterbury Tales* 1191-1624).

5. In Le Maçon, *Les Dix Journées de Jean Boccace*, a translation utilized by La Fontaine, the widow "[avait] faict faire secretement un baing, et apprester magnificquement à soupper" (II, 19).

CHAPTER SEVEN

CASUISTRY AND THE FANTASY OF GUILTLESS SEX

Although La Fontaine took part seriously in the fierce debates between Jansenists and Jesuits on the merits of the casuistic approach to moral questions, in his *Contes* casuistry becomes no more than another playfully ironic component of the game of love. For the moral theologian, casuistry was a legitimate and necessary application of general moral principles to practical cases. For the parish priest the penitential handbooks, the *institutiones morales*, and the *resolutiones morales* (Cariou 119-22), provided useful advice on judging the gravity of sins and on giving advice to penitents.[1] The danger of being either too rigorous or too indulgent in solving the dilemmas presented by conflicting moral obligations was always present and provided wonderful opportunities for satiric attack and comic parody.

La Fontaine's closest relationship with the Jansenists of Port Royal was during the period of 1664-74, the decade when he published the *Contes*.[2] The ironic side to this coincidence did not escape Pierre Clarac: "Que l'auteur des *Contes*, au moment même où ils paraissaient, entretienne des relations suivies avec les Messieurs de Port Royal, on s'en étonnera ou on s'en amusera: en tout cas, la chose n'est pas douteuse" ("La Fontaine," 240). The Jansenists were preoccupied with defending Christian morality against a spirit of minimalism and laxism that they attributed to Jesuit casuistry. Not content with attacking the excesses of casuistry, they launched an all-out war, enlisting the aid of Blaise Pascal. His *Provinciales* (1656-57) brought notoriety to the chief victims of his attack, the Jesuits Escobar and Étienne Bauny.[3] About the year 1664 La Fontaine composed a "Ballade sur Escobar" and another poem entitled "Stances sur le même sujet."[4] These poems circulated in 1666, but were not published until sometime after La Fontaine's death in 1695. The refrain of his ballad became so well-known that Richelet's *Dictionnaire* cited it as early as 1680 under the entry for *velours*: "Escobar sait un chemin de velours." In these two satiric poems, La

Fontaine took up the Jansenists' arguments against the moral laxity of the Jesuits, and in particular of Escobar:

> Il ne dit pas qu'on peut tuer un homme
> Qui sans raison nous tient en altercas,
> Pour un fétu, mais bien pour une pomme,
> Ou tout au moins pour quelque ducats.
> Même il soutient qu'on peut, en certains cas,
> Faire un serment plein de supercherie,
> S'abandonner aux douceurs de la vie,
> S'il est besoin, conserver ses amours,
> Ne faut-il pas après que l'on s'écrie;
> Escobar sait un chemin de velours"?
>
> (ll. 11-20)[5]

In the *Contes*, however, casuistry manifests itself as only another source of *badinage* with its comic juggling of words, its search for euphemistic detours of language and witty pseudoarguments that serve to salve uneasy consciences. "La Fiancée du roi de Garbe" constitutes La Fontaine's most elaborate exemplum of self-delusional casuistry that attempts to provide a legitimization for morally forbidden acts. Reduced to a form of wordplay in "Le Cas de conscience," to a *question galante* in "L'Abbesse," casuistry, in the end, like lying, becomes a defensive strategy for some ("Le Calendrier des vieillards") and an offensive strategy for others ("Les Frères de Catalogne," "Pâté d'anguille"). Casuistry generates deceit structures that fall into two distinct patterns of false interpretive framing. In the first ("La Fiancée du roi de Garbe," "L'Abbesse"), casuistry provides a transforming frame that makes the sex act appear virtuous, or at least blameless. In the second, one moral case is juxtaposed to another providing a frame of interpretation that applies duplicitously to a completely different set of circumstances. This is the deceit pattern in "Le Cas de conscience," "Les Frères de Catalogne," "Pâté d'anguille," or "Le Calendrier des vieillards."

"La Fiancée du roi de Garbe"[6] was published in 1666, the same year that La Fontaine's poems on Escobar were circulating. This long poem (801 lines) owes little to Boccaccio: "Jamais encore La Fontaine conteur n'a revendiqué liberté plus entière dans l'imitation;...il ne garde en somme qu'un cadre vide, qu'il remplit à sa fantaisie" (Collinet 1399).[7] Alaciel, the young heroine, has to make her way from the court of her father, Zaïr, in Alexandria, to the court of her future husband, Mamolin, king of Garbe. During her passage from one patriarchal territory to the other, she makes an involuntary detour, accumulating successively along the way eight lovers, before being united in the end with her unsuspecting fiancé. The "story" is encapsulated in line 31: "Veuve de huit galants, il la prit pour pucelle." Throughout the ensuing narrative expansion, the narrator amuses himself by presenting the workings of Alaciel's own naive casuistry and by interpreting her actions to the reader in a casuistic light. Her tool of self-deception is casuistry.

From the outset La Fontaine's text accentuates by repetition the word *accommodement*, a key word that thematizes casuistry in the tale. In the debates over casuistry *accommodement* became a semantically charged word, whose very mention conjured up the specious arguments of casuistry. These connotations are evoked in the famous passage from Act Four of Molière's *Tartuffe* (1664) where Tartuffe provides a satiric portrait of casuistic methodology:

> Le Ciel défend, de vrai, certains contentements;
> Mais on trouve avec lui des accommodements;
> Selon divers besoins, il est une science
> D'étendre les liens de la conscience,
> Et de rectifier le mal de l'action
> Avec la pureté de notre intention.
> (ll. 1487-92)

Alaciel is constantly called upon to adapt to new situations, to stretch the boundaries of her conscience—"selon divers besoins"—without really transgressing the limits of acceptable moral conduct. Early in the tale Alaciel is described as being, above all, "accommodant": "De bon, de beau, de charmant, et d'aimable, / D'accommodant, j'y mets encor ce point" (l. 43). About halfway through the text, the infinitive *s'accommoder* occurs to describe Alaciel's conduct: "Qu'eût fait Alaciel? force n'a point de loi. / S'accommoder à tout est chose nécessaire" (ll. 385-86). Then, again, near the conclusion, the same leitmotif recurs as the narrator praises the heroine who is finally united with her fiancé: "Et je puis dire à son honneur / Que de tout elle s'accommode" (ll. 764-65). Her manner of acting is worthy of those casuists castigated by Pascal in the *Provinciales* (1656): "Vous avez entrepris de décider des cas de conscience d'une manière favorable et accommodante" (810). Alaciel is full of "commodités" (l. 407). This calls to mind how Madame de Montespan, as well, played with the multiple meanings of this word calling the casuistic confessor of Louis XIV, le père La Chaise, "une chaise de commodités."[8]

Alaciel's adventures begin when the giant Grifonio and his pirates take over her ship and she falls into the sea, accompanied by her beloved companion, Hispal. This fall, opening as it does the parenthesis of her amorous adventures, is like many another literary *chute* the symbol of a psychopathological fall to come: "La chute," according to Gilbert Durand, "devient alors l'emblème des péchés de fornication" (111). But, as the casuistic narrator urges the reader from the outset, Alaciel is not to be judged harshly, for throughout her sexual encounters her intentions remain pure:

> Il ne faut pas pour cela qu'on l'accuse:
> Ce n'était après tout que bonne intention,
> Gratitude, ou compassion,
> Crainte de pis, honnête excuse.
> (ll. 26-29)

One of the fundamental moral principles of the casuists was, after all, the direction of one's intention to a good end, as Pascal mockingly explained: "Quand nous ne pouvons pas empêcher l'action, nous purifions au moins l'intention; et ainsi nous corrigeons le vice du moyen par la pureté de la fin" (729).

Hispal becomes her first lover, primarily because she is grateful to him:

> Les services d'Hispal en ce même moment
> Lui reviennent devant la vue.
> Ses jours sauvés des flots, son honneur d'un géant.
> (ll. 251-53)

A further argument that persuades her to enter the cave (another sexual "emblem" for Durand!) is the assurance of complete secrecy. Estienne Bauny, after all, assures his readers that a sin is less grave when there is no scandal attached to it: "quand l'acte de luxure a esté commis si secrettement, qu'on n'en a point de connoissance, car qui se scandaliseroit de ce qu'il ne voit pas?" (125) Tartuffe's famous line comes to mind: "Et ce n'est pas pécher que pécher en silence" (l. 1566).

Next, an anonymous captain, about to starve himself to death on Alaciel's account, is saved by her charitable compassion:

> Laisser mourir un homme, et pouvoir l'empêcher!
> C'est avoir l'âme un peu trop dure.
> Par pitié donc elle condescendit
> Aux volontés du capitaine.
> (ll. 344-47)

The third "galant" is a pirate who captures her and coerces her into yielding to him by depriving her of food. What could she do?

Qu'eût fait Alaciel? force n'a point de loi.
S'accommoder à tout est chose nécessaire.
Ce qu'on ne voudrait pas, souvent il le faut faire.
(ll. 385-87)

As the casuist Bauny pointed out, necessity legitimizes many an undesirable, even illicit, action: "*Quod non est licitum in lege, necessitas facit licitum*" (712).

In the fourth letter of the *Provinciales*, the casuist invokes surprise as an attenuating circumstance for judging the sinfulness of an action: "Tous les péchés de surprise ne pourraient être imputés, puisque avant de les commettre on n'a ni la connaissance du mal qui y est, ni la volonté de l'éviter" (693). The next escapade of Alaciel is a sin of surprise. The fourth lover, having carried her off to his château and plied her with wine—which she, naturally, had never tasted before—attacks her while she is sleeping:

Bacchus donc, et Morphée, et l'hôte de la belle,
Cette nuit disposèrent d'elle.
Les charmes des premiers dissipés à la fin,
La princesse au sortir du somme
Se trouva dans les bras d'un homme.
(ll. 491-95)

Without her knowledge, this lover departs and leaves her in the care of another man who secretly crawls into her bed and is the occasion of another "péché de surprise." After all, she thought she was just continuing to make love with the *seigneur de château*, which would not count as a new sin. According to Bauny, a second sin would only occur if the internal consent and the action itself were interrupted: "l'une et l'autre action conjoinctement doivent estre interrompues, continuer en l'une d'elles, c'est assez à ce que le péché soit un, et non plusieurs en nombre" (668). Was this classical argument not what Alaciel's last lover had counselled?

Une nuit, dit-il, est de même que cent,
Ce n'est que la première à quoi l'on trouve à dire.
(ll. 500-501)

The sixth sexual transgression was motivated by the charitable intention of saving a young girl who was about to be raped:

Il ne sera pas dit que l'on ait, moi présente,
Violenté cette innocente.
Je me résous plutôt à toute extrémité.
Ce combat plein de charité
Fut par le sort à la fin terminé.
L'infante en eut toute la gloire.
(ll. 598-603)

With the seventh and eighth lovers the argument of necessity again excuses her misdemeanors. She is all alone and a chevalier proposes to take her back to her father's court in consideration for certain favors:

Vous demeureriez seule, et pour vous parlez franc
Je tiens ce service assez grand,
Pour me flatter d'une espérance
De récompense.
(ll. 668-71)

Alaciel agrees, but only accepts to fulfill her duties toward him within the terms of a carefully worded contract:

Avec beaucoup de foi le traité s'exécute:
Pas la moindre ombre de dispute:
Point de faute au calcul, non plus qu'entre marchands,
De faveur en faveur (ainsi comptaient ces gens)
Jusqu'au bord de la mer enfin ils arrivèrent,
Et s'embarquèrent.
(ll. 692-97)

Then, just before he dies, her companion asks his nephew to care for Alaciel and inherit "le surplus des faveurs" (l. 715). Her last payment takes place right on the border of her father's land, "justement sur le bord / De la frontière" (ll. 721-22). The parenthesis of her amorous adventures is closed.

Protected by the self-delusionary charm of casuistry, she has traversed intact the dubious never-never land between the domain of her father and the domain of her new husband. And, paradoxically, at the end of "La Fiancée du roi de Garbe" the same casuistry of love that brought her through so many dangers, now stands her in good stead as she enters the new territory of her marriage bed:

Alaciel conta tout ce qu'elle voulut.
Dit les mensonges qu'il lui plut.
Mamolin et sa cour écoutaient en silence.
La nuit vint: on porta la reine dans son lit.
A son honneur elle en sortit:
Le prince en rendit témoignage.
(ll. 773-78)

"Le Cas de conscience"[9] is a delightful parody of the casuistic method of drawing a precept from one case and applying it *ad absurdum* to another situation. Quite likely of La Fontaine's invention, the narrative presents Anne as a young woman with a scrupulous conscience—Escobar describes scrupulosity as a tendency to find sin even where there is none: "*vana apprehensio, et hinc ortus timor peccatum alicubi esse, ubi non est* (17-18). Her sin consists in having lingeringly observed a nude young man bathing in the river. Her guilt would be minimal, even according to the most rigorous seventeenth-century theologians, because she did not at the time believe the offense to be serious:

Le scrupule survint, et pensa tout gâter.
Anne avait bonne conscience:
Mais comment s'abstenir? est-il quelque défense
Qui l'emporte sur le désir
Quand le hasard fait naître un sujet de plaisir?
La belle à celui-ci fit quelque résistance.
A la fin ne comprenant pas
Comme on peut pécher de cent pas,
Elle s'assit sur l'herbe; et très fort attentive
Annette la contemplative
Regarda de son mieux...
(ll. 55-65)

One of her obligations as a Catholic was to go to confession at Eastertime, so Anne duly presents her sins in the confessional, but tries to gloss over the one she is most ashamed of:

Pâques vint, et ce fut un nouvel embarras.
Anne faisant passer ses péchés en revue
Comme un passe-volant, mit en un coin ce cas;
Mais la chose fut aperçue.
(ll. 86-89)

Her confessor knows that his duty is to ask for more details about the case so that he can first judge whether it is a mortal sin or a venial sin, and then give her an appropriate penance:

> En confesseur exact il fit conter l'histoire,
> Et circonstancier le tout fort amplement,
> Pour en connaître l'importance,
> Puis faire aucunement cadrer la pénitence,
> Chose où ne doit errer un confesseur prudent.
> (ll. 92-96)

Now, according to Estienne Bauny's manual for confessors, chapter 9 entitled "Des choses dont il faut interroger le Pénitent en fait de luxure," Anne could be considered guilty of serious sin:

> S'il a fait des regards impudiques sur soy, ou d'autres avec lubricité, pesché de même qualité, et qui est mortel en soy; puis qu'au tesmoignage du Fils de Dieu, voir une femme pour la convoiter, c'est avoir ja commis l'adultere en son coeur, *Qui videris mulierem ad concupiscendum eam, iam moechatus est eam in corde suo.* (122)

And yet, Anne, after all, was only admiring the beauty of one of God's creatures—"au corps jeune et frais, / Blanc, poli, bien formé, de taille haute et drète" (ll. 49-50). Bauny, himself, might have understood, for, as he added to his commentary: "Toutefois s'arrester sur la beauté desdites créatures, et s'y plaire, sans passer plus avant, n'est au plus que péché véniel" (122). Anne's confessor was more severe:

> Etre dans ses regards à ce point sensuelle!
> C'est, dit-il, un très grand péché;
> Autant vaut l'avoir vu que de l'avoir touché.
> (ll. 98-100)

As part of the penance enjoined on her, she one day brings her confessor an enormous freshly caught fish. Since he is about to host a dinner party, he asks her to cook the fish and bring it back later for dinner. Anne, as an amateur casuist, wreaks her revenge by returning without the fish, but applying the priest's casuistical precept to his own case: "Autant vaut l'avoir vu que de l'avoir mangé" (l. 152). The casuistry of Anne operates as an agent of dupery precisely because it is a nonfunctional space-builder; as a parody it points up the weakness of a logical system that utilizes general casuistic precepts as interpretive frames for a diversity of cases.

The nun in "L'Abbesse," like Alaciel in "La Fiancée du roi de Garbe," is a representation of casuistry functioning as self-deception. She finds a way to justify sexual activity without violating her vow of chastity. Like many a *question galante*, the dilemma posed in this tale is a forced one, couched as it already is from the outset in impossible terms, a result of wishful thinking by both storyteller and imaginary heroine. Casuists were regularly accused of laxism, of using their specious reasoning to stretch the boundaries of moral law too far. The abbess's learned physicians decide that the only remedy for her deathly illness is to have "compagnie d'homme" (l. 67). The mischievous nature of the remedy proposed for her mysterious illness reminds one of some of the more absurdly impossible questions posed by the casuists, such as, how long could a married couple locked in a church be there without sinning against either the sanctity of the place by having sexual intercourse or against the sacred duties of marriage by not doing so. As Martha M. Houle demonstrates, in "L'Abbesse" the focus is not on the moral problem, but on external appearances, the *question galante*: "La Fontaine fuses two worlds in his tale where the abbess poses a case of conscience not as a moral problem,...but as a worldly issue, as a *question galante* ("The Play of *bienséances*" 112). How can the abbess find a way to present herself to her own community of nuns as someone who approves of such a remedy?

> ....non pas que dans son âme
> Ce bon ne fût par elle souhaité:
> Mais le moyen que sa communauté
> Lui vît sans peine approuver telle chose?
> (ll. 77-80)

Her *souhait* is a mirror of the unacknowledged and unmentionable desires of the other nuns. Her medical prescription provides them as well with the protective cloak of casuistry, a means of hiding both from themselves and from others the reality of their forbidden desires. After all, in the logic of casuistry the good intention behind an action justifies the action itself, making it blameless. That is why when Soeur Agnès presents the situation to the other nuns, she starts with the exculpating intention, "par charité":

> Par charité n'en est-il point quelqu'une
> Pour lui montrer l'exemple et le chemin?
> (ll. 104-5)

Seen through the cognitive frame provided by casuistry, sexual transgression is interpreted only too willingly as a virtuous action. Like the Rabelaisian flock of sheep in La Fontaine's prologue to "L'Abbesse,"[10] all of the nuns hurtle past the sexual taboo into the arms of the first man they can find—"des noirs, et des blancs, / Et des tannés" (ll. 112-13):

> ....Il n'est aucune ouaille
> Qui ne s'y jette; ainsi que les moutons
> De Dindenaut dont tantôt nous parlions
> S'allaient jeter chez la gent porte-écaille.
> (ll. 126-29)

The abbess, at last, finds her *remède* in an explosion of colors—"Elle redevient rose, / Oeillet, aurore" (ll. 134-35)—and then the author-narrator translates her satisfaction through the lyric style of the final lines. The abbess's conscience has in the end been lulled into silence, perhaps not so much by the arguments of casuistry as by the contentment of sexual pleasure expressed in the final anaphoric outburst:

> O doux remède, ô remède à donner,
> Remède ami de mainte créature,
> Ami des gens, ami de la nature,
> Ami de tout, point d'honneur excepté.
> Point d'honneur est une autre maladie:
> Dans ses écrits madame Faculté
> N'en parle point. Que de maux en la vie!
> (ll. 137-43)

For the monks in "Les Frères de Catalogne," stretching the definition of tithing, *la dîme*, to include demanding sexual favors from the local wives is a logical maneuver. If the parishioners pay a tithe on all their goods, why should there not be a tithe on "les oeuvres de mariage" (l. 66)? Their deformation of casuistic logic becomes the occasion for a witty poetic passage where the insistent rhythm of the *rimes plates* at the end of brief octosyllabic lines gives the force of formulaic legal prescriptions to the specious arguments:

> Rendez grâce aux bontés célestes,
> Nous laissant dîmer, sur un bien
> Qui ne vous coûte presque rien.
> C'est un droit, ô troupe fidèle,

Qui vous témoigne notre zèle;
Droit authentique et bien signé,
Que les papes nous ont donné;
Droit enfin, et non pas aumône:
Toute femme doit en personne
S'en acquitter trois fois le mois,
Vers les frères catalanois.
Cela fondé sur l'Ecriture,
Car il n'est bien dans la nature,
(Je le répète, écoutez-moi)
Qui ne subisse cette loi
De reconnaissance et d'hommage:
Or les oeuvres de mariage
Etant un bien, comme savez,
Ou savoir chacune devez,
Il est clair que dîme en est due.
(ll. 50-69)[11]

The women accept the false *savoir* and their prescribed duty with a casuistic *dévotion* (l. 112); their husbands, when they discover the subterfuge, are less impressed with such religious zeal. With the whole village enraged by their conduct, the monks, like so many tomes of outdated casuistry, are burned in a joyous village fire:

Les maris sautants alentour,
Et dansants au son du tambour.
Rien n'échappa de leur colère,
Ni moinillon, ni béat père.
Robes, manteaux, et cocluchons,
Tout fut brûlé comme cochons.
(ll. 251-56)

A pseudologic is played with as well in "Paté d'Anguille" where a husband whose guiding precept in love is "Diversité, c'est ma devise," uses a false analogy to convince his reluctant valet to share his wife with his master. Every day he has a delicious "paté d'anguille" served to the valet, until one day the valet rebels and wants to share his master's plate:

M'en voilà soûl, reprit le sire.
Et quoi toujours pâtés au bec!

Pas une anguille de rôtie!
Pâtés tous les jours de ma vie!
J'aimerais mieux du pain tout sec:
Laissez-moi prendre un peu du vôtre.
(ll. 74-79)

The master draws a specious analogy with his own desire to sleep with the valet's wife:

Qu'ai-je fait qui fût plus étrange?
Vous me blâmez lorsque je change
Un mets que vous croyez friand,
Et vous en faites tout autant.
(ll. 93-96)

From the prose narrative of *Les Cent Nouvelles nouvelles* (10) La Fontaine has fashioned a sprightly poem in octosyllables, tightly unified by the narrator's periodic refrain "Diversité, c'est ma devise" (ll. 4, 13, 17, 101, 147) and by the metaphoric network of references to food. Where the original tale simply commented at the end: "passa le temps comme il souloit avecques les belles et bonnes" (*Conteurs français* 57), La Fontaine's concludes "Paté d'Anguille" in the manner of a male sexual fantasy with an orgiastic enumeration of the women this man has consumed :

Par où le drôle en put croquer,
Il en croqua, femmes et filles,
Nymphes, grisettes, ce qu'il put.
Toutes étaient de bonne prise;
Et sur ce point, tant qu'il vécut,
Diversité fut sa devise.
(ll. 142-47)

In "Le Calendrier des vieillards," a tale adapted from the *Decameron* (II, 10), religious discourse becomes the defensive weapon that Richard, the elderly husband of Bartholomée, cleverly wields to mask his reluctance to make love with his wife. Richard, well-versed in the interpretation of law (l. 24), applies his talent in casuistic fashion to his own marital dilemma, determining that on his religious calendar almost every day, for one reason or another, is a day for abstaining from lovemaking:

Toute en féries il mettait la semaine;
Et bien souvent faisait venir en jeu
Saint qui ne fut jamais dans la *Légende*.
Le vendredi, disait-il nous demande
D'autres pensers, ainsi que chacun sait:
Pareillement il faut que l'on retranche
Le samedi, non sans juste sujet,
D'autant que c'est la veille du dimanche.
Pour ce dernier, c'est un jour de repos.
Quant au lundi, je ne trouve à propos
De commencer par ce point la semaine:
Ce n'est le fait d'une âme bien chrétienne.
Les autres jours autrement s'excusait.
(ll. 49-61)

The result of this casuistry is that Bartholomée remains unconvinced, and happily accepts to be unfaithful to her husband when Pagamin, a well-mannered pirate, carries her off. Her belief frame has been unchanged by Richard's feigned scruples. His casuistry actually has the unintended effect of inciting her to relinquish gladly her husband and his calendar for a new way of viewing sexuality, without rules:

Mais Pagamin pour tous y pourvoira.
Il ne sait loi, ni digeste, ni code;
Et cependant très bonne est sa méthode.
. . . .
Adieu vous dis, vous et vos jours de fête.
(ll. 229-31, 238)

***

In the introductory lines of "L'Oraison de saint Julien" the narrator comments on the power of words:

Bien est-il vrai qu'auprès d'une beauté
Paroles ont des vertus nonpareilles;
Paroles font en amour des merveilles:
Tout coeur se laisse à ce charme amollir.
(ll. 6-9)

The casuistry of love does not operate on the ground of ideology. Like lies and false promises, it is a form of verbal manipulation of cognitive space. The casuistic framing of moral situations, when it is abused, leads to the construction of self-delusional false belief spaces. Casuistic language and argumentation generate in the literary text of the *Contes* a playful cognitive pseudoterritory of guiltless sex for the diegetic characters, while on the metanarrative terrain reader and narrator disport themselves among the cognitive gaps and the discrepant knowledge frames.

NOTES

1. On casuistry, see the article, "Casuïstique," by Louis-Gustave Vareecke; and Albert R. Jonsen and Stephen Toulmin, *The Abuse of Casuistry*. During the sixteenth and seventeenth centuries manuals containing compilations of cases, *summae casuum*, were numerous, more than 600 between 1560-1660. The *Summa resolutionum moralium* of Diana presented more than 20,000 cases.

2. In 1665 and 1667 the Jansenists of Port Royal had Augustine's *City of God* translated by Louis Giry and published in 2 volumes by Le Petit, their favorite editor; La Fontaine did a verse translation of the poetic passages. A few years later the Jansenists asked him to write a dedicatory poem for their *Recueil de poésies chrestiennes* (1671). Also, under their instigation La Fontaine published the *Poème sur la captivité de Saint Malc* in 1673; ironically, this praise of Christian virginity appeared only a few months before the publication of the last volume of the *Contes*.

3. Anthonius Escobar, *Liber theologiae moralis* and Estienne Bauny, *La Somme des péchés*. On Pascal and casuistry, see Gérard Ferreyrolles, *Blaise Pascal: Les Provinciales* and Pierre Cariou, *Pascal et la casuïstique*. More particularly on La Fontaine's relationship with the Jesuits, see Pierre Clarac, *La Fontaine par lui-même*.

4. La Fontaine, *Oeuvres diverses*, ed. Pierre Clarac, 587-90. The poems are found together in the Bibliothèque Nationale Ms. fr. 12618, p. 5, the first bearing the date 1666.

5. Jean Orieux, *La Fontaine* uses the ballad's refrain to highlight the apparent inconsistency between La Fontaine's epicurean temperament and his connections with the Jansenists: "Que va faire dans cette triste galère notre épicurien, cet ami du sommeil, de l'amour, de la musique, de la table, du vin, et des fêtes de

Saint-Mandé ou de Vaux?... S'il est homme au monde qui a choisi 'le chemin de velours', c'est bien La Fontaine" (168).

6. See my article, "Erotic Dimensions of Space in La Fontaine's 'La Fiancée du roi de Garbe" and Jane Merino-Morais's chapter 2, "La Fiancée du roi de Garbe: la circulation de l'objet de valeur"; Manfred Gsteiger compares the narrative technique in La Fontaine's tale with Boccaccio's, although the tales are so different that such a method renders dubious results.

7. A puzzling castigation of La Fontaine's tale as a poor translation that has harmed the reputation of Boccaccio has surfaced once in a while, ever since that opinion was expressed in 1863 by Emile Montagut: "Notre bon La Fontaine, pour comble de malheur, a pris cette histoire dans Boccace et en a fait un de ces contes lestes, grivois, qu'il fait si bien, en sorte que la seule fiancée du roi de Garbe que l'on connaisse est celle de La Fontaine et non pas celle de Boccace. Le conte gai et grivois a fait oublier le conte sérieux et dramatique. La plupart des lecteurs ont cru que la traduction de notre poète pouvait les dispenser de l'original, et ils ne savent pas que cette traduction est infidèle" (731). Gérard Genot, in his article on "La Fontaine et Boccace," speaks of "la captation presque totale d'un auteur par une autre littérature" (579).

8. See Françoise Chandernagor, *L'Allée du roi*, 307.

9. See the article by Martha M. Houle, "La Fontaine's Fun with Casuistry in *Le Cas de conscience*." Some similarity has been noted between this tale and a piece in *Le Facétieux Réveille-matin*—see Jacqueline Plantié, "L'Aboutissement de la tradition facétieuse dans une fable et un conte de La Fontaine."

10. Anne L. Birberick finds in the analogy between sheep and nuns a key to the interpretation of "L'Abbesse" (186-90). She deftly analyzes the recurring analogy for its function in the satire of religious mores and its metafictional significance.

11. In La Fontaine's source, *Les Cent Nouvelles nouvelles* (32), this whole argumentative development is absent.

CONCLUSION

La Fontaine's practice of highlighting divergent cognitive perspectives is a significant feature of his entire literary production. In the *Fables*, one major source of multiple perspectives for the reader is the allegorical genre itself. The fable, whose narrative has its ostensible locus in the animal world, is ambiguous in that it must simultaneously be interpreted in human terms. Patrick Dandrey characterizes the poetics of the *Fables* as being based on the rhetorical figure of syllepsis, "figure de l'ambiguïté s'il en est, puisque désignant le jeu subtil du recouvrement partiel de deux sens, l'un propre et l'autre figuré, qui par superposition et transparence s'imposent ensemble à l'esprit" (215). The *Contes* generate in a different way a kaleidoscope of cognitive perspectives. Both works reflect the typically seventeenth-century literary practice of presenting a multiplicity of views on the same reality.[1] The dilemmas of Rodrigue and Chimène in *Le Cid*, the *questions d'amour* of the précieuses, Tartuffe's casuistry, the "love questions" in the *Princesse de Clèves*, the *équivoques* and irony of Racine's theatre, and even Pascal's wager—these are all examples of constructing double visions of a single reality. Perhaps the source of this underlying episteme can be traced to a nascent relativism, reflected in Montaigne's skeptical turn of mind: "Quand je me jouë à ma chatte, qui sçait si elle passe son temps de moy plus que je ne fay d'elle?...Elles nous peuvent estimer bestes, comme nous les en estimons" (*Essais* I, 12: 496-97). Montaigne had also delved into the diversity of moral customs practiced among the cannibals (*Essais* I, 31). And Pascal had pondered: "Vérité au-deçà des Pyrénées, erreur au-delà" (*Pensées* 1149). In the *Fables*, declares Dandrey, "les mêmes vers délivrent l'un et l'autre propos selon l'angle sous lequel on les perçoit" (216). In the *Contes* generating dilemmas and juggling discrepant states of belief or knowledge is an essential part of La Fontaine's literary agenda.

The dynamic cognitive framing at work within the storyworld of the *Contes* has been the main focus throughout this study. My aim is to more fully understand the often complex patterns operating in La Fontaine's tales by coupling a method of frame analysis with a sensitivity to the poetic texture of the *Contes*. Frame analysis is, I believe, particularly suited to the task of accounting for the multiple and sometimes convoluted cognitive layering of fictional texts. The elaboration of a theory of frame analysis appropriate for application to literature has, however, only begun, and it is not a simple project. By studying the functioning of deceit in La Fontaine's text this study proposes one approach through frame analysis and points to directions in which further adaptations to other textual studies might be envisaged.

My approach is to some extent taxonomic in that it offers a new way of dividing the tales into categories. Based on the factors that effect the deceit—lying, false promises, magic, disguise, displacement, casuistry—this is a classification aimed less at establishing a repertoire of static motifs, as the folklorists are wont to do, than on finding patterns in the dynamic narrative forces operating in the storyworld. Instead of simply noting, for example, that the magic pear tree motif occurs in "La Gageure des trois commères" (Aarne and Thompson, Tale Type 1423), I examine how the magical illusion creates multiple cognitive perspectives and how the generation of these false frames brings about deception.

To be sure, the narrative kernels of the *contes* can be linked to traditional folkloric motifs and types of the folktale. The folklore scholar G. Legman sees the *conte en vers* as an important embodiment of folktale material:

> The *conte-en-vers* is therefore the real connecting link between the earlier folktale collections and the modern scientific collections, and is the real repository of the native and original humorous material for nearly two centuries (1650-1850), where little of an original nature and almost nothing native will be found in the jestbooks. The entire literature of the *conte-en-vers*, whether dated from the burlesque and Aretinesque academies of the 1530's in Italy, or from La Fontaine's *contes*, "tirés de Boccace," as he perfectly frankly admits, in the 1660's, or even, for that matter, from the exactly similar method of the folktale-based fabliaux centuries before, represents in sum an exceptionally large repertory of jokes and tales, purposely *sought from the folk* at a period when the jestbooks were already forgetful or contemptuous of folk sources, and far gone in sterile mutual plagiarism. (237)

Rather than choosing to follow the folklorists' methodology for categorizing tales, it seems more appropriate for a literary study to highlight the narrative "transformations," as Greimas—among others—termed them,[2] by categorizing the tales in the *Contes* according to their dominant operative act of deception. Even among folklorists there is controversy about how tales should be classified; traditionalists continue to classify tales according to professions, activities, character types, classes, and nationalities, whereas Legman, for example, urges reclassification of erotic folktales, much as I have done, according to "the operative point of view, by the sexual nexus on which the story turns—that essential element without which it cannot exist" (238).[3]

At the conclusion of this detailed analysis of all of La Fontaine's *contes en vers*, we are now in a position to look for a more synthetic view of the patterns of deceit that have emerged. One obvious conclusion is that the instruments of deceit, those operative acts such as lying or false promises, tend to generate cognitive spaces of a particular nature. The false promise produces hypothetical frames; the disguise produces counterfactual spaces; magic and pseudomagic often produce doubtful spaces. The nature of the spaces and of their frames—counterfactual, doubtful, or hypothetical—generated by deceit have been crucial to the analysis of individual tales. Of more significance, however, from the taxonomic point of view are the actual patterns of deceit fashioned by the cognitive frames. Once the preliminary classification by the operators of deceit is done, the patterns of deceit afford a further means of comparing the diverse configurations of cognitive space in these narratives.

There appear to be three basic patterns of deceit operative in the *Contes*: the simple deceit frame, embedded frames, and juxtaposed frames. The first type is the simple dupery accomplished by a deceiver who falsifies the belief of a deceivee through a disguise, a lie, or some other instrument of deception and so achieves the goal, which is usually sexual in nature. A single false belief space is generated in this type of deception. An example of the simple deceit frame occurs in "Le Cuvier" when a woman lies to her husband in order to conceal the presence of her lover.

The Simple Deceit Frame

Observer's Frame

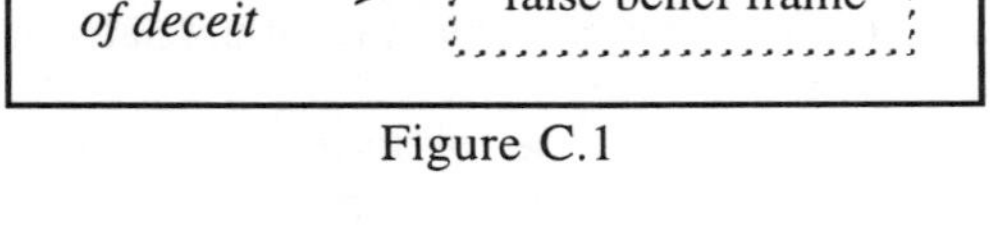

Figure C.1

The second cognitive pattern of deceit comprises embedded frames. Here more than one cognitive frame is present and several levels of deception operate simultaneously. Frequently the tale presents a *trompé se croyant trompeur* narrative, as for example, in "Le Cocu battu et content" in which Monsieur Bon believes that his disguise is duping his wife's lover, while, in fact, he is the dupe who is being manipulated by the disguise.

Embedded Frames

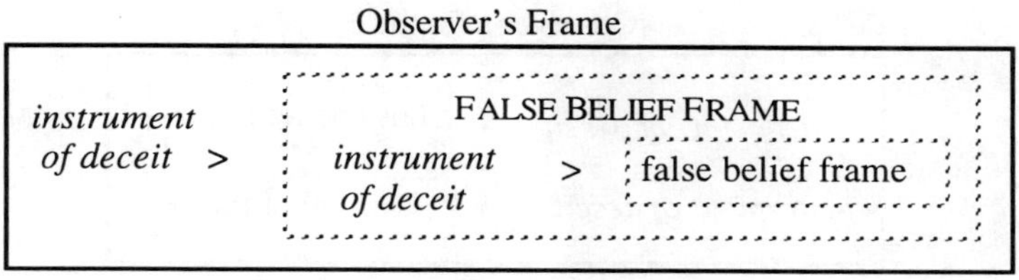

Figure C.2

The deception generated by a false promise is a special variety of an embedded pattern that is less complex than the pattern in the preceding schema because the deceivee does not produce the false vision of another character's counterfactual belief. The embedded frame is situated within the deceivee's own false belief space; it is a perception of false hope, a cognitive space that is hypothetical and anticipatory in nature. When the deceiver in "Comment l'esprit vient aux filles" promises Lise that she will become intelligent if she sleeps with him, that false promise evokes such a strong desire to obtain the anticipated goal that she consents to his sexual advances.

Embedded Anticipation Frames

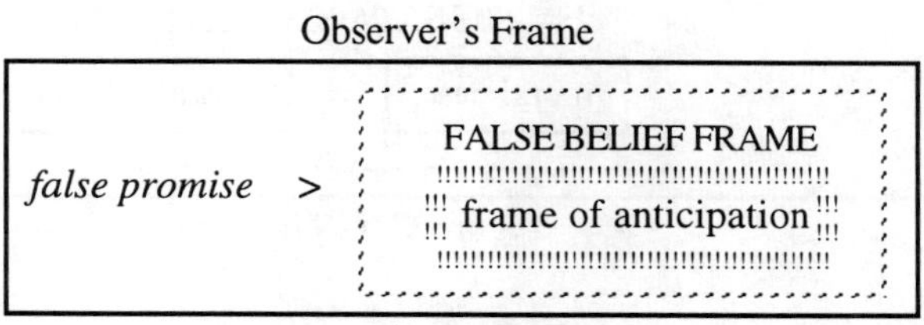

Figure C.3

The third basic type of deceit pattern, juxtaposed frames, displays two or more cognitive frames linked to each other, most often in a causal relationship. Such is the pattern of "La Mandragore" in which both the husband, Nicia, and his wife, Lucrèce, must be duped successively; the *trompeur trompé* narratives—like "Le Petit Chien qui secoue de l'argent et des pierreries"—also fit into this category.

Juxtaposed Frames

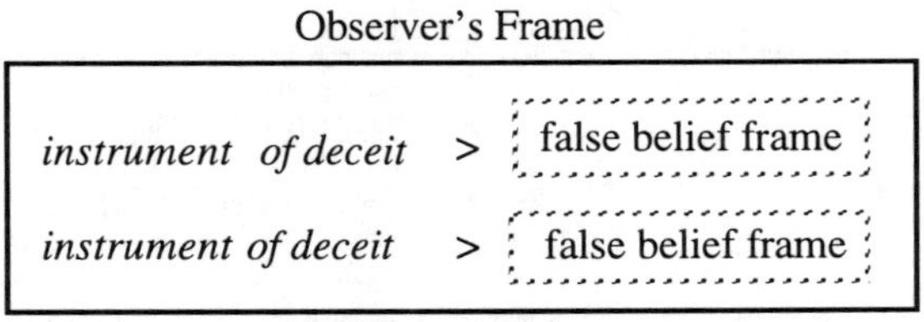

Figure C.4

Deceit patterns generated by casuistry deserve special attention. When the duplicity consists in presenting two moral cases or a precept of the law and its practical application, one juxtaposed frame aids in the interpretation of the other. In "Le Cas de conscience" when Anne arrives at the curé's dinner party without her fish, her pretext is that from the confessor's framing of her own earlier misdemeanor ("Autant vaut l'avoir vu que de l'avoir touché") she derives an application to her present situation: "Autant vaut l'avoir vu que de l'avoir mangé."

Juxtaposed Casuistical Frames

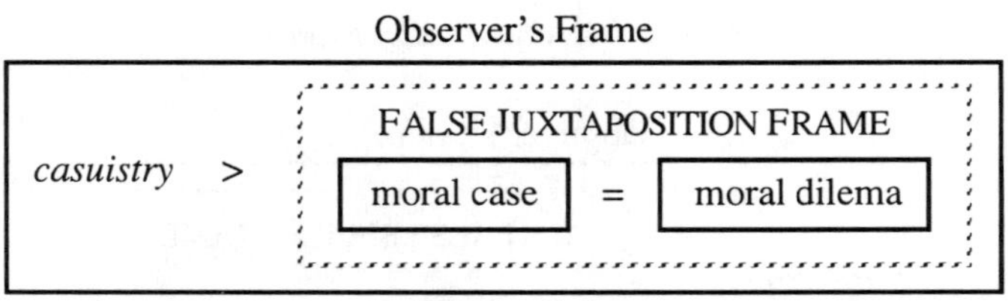

Figure C.5

The other pattern which may be constructed by casuistry is a simple deceit pattern in which the casuistical reasoning produces a false perspective through which to interpret the moral dilemma. A forbidden sexual activity may be viewed as permissible if—as in the case of Alaciel in "La Fiancée du roi de Garbe"—it is directed toward a charitable end, or if it is motivated by the excusing factor of circumstances, such as secrecy, surprise, or necessity.

Simple Casuistical Frame

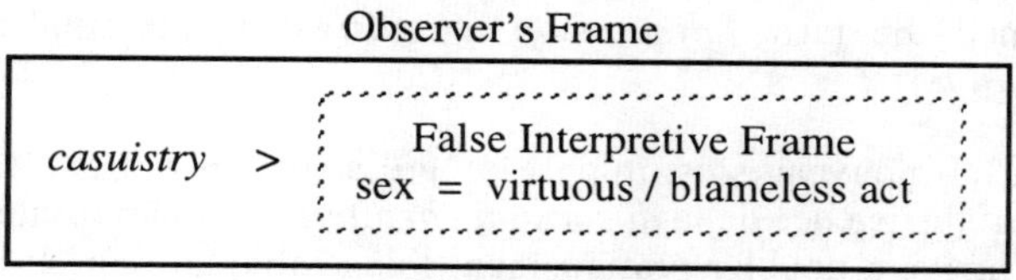

Figure C.6

A further cognitive complexity I discovered in my analysis is that in a good number of tales unintentional deceit occurs. When the deceit is unintentional, the deceiver does not function as an all-knowing observer. "Les Troqueurs" is an example of naive embedded frames; "Les Quiproquos" is an example of naive juxtaposed frames.

Many *contes* present more than one act of deception. How do multiple acts of deception relate functionally to each other in the same tale? Just as a well constructed drama has only one culminating point, each tale normally has only one climactic act of deception. In a few instances, however, serial deceit patterns of a similar type can be discerned. "La Fiancée du roi de Garbe," for example, embodies a series of juxtaposed casuistical frames. Other tales, like "Le Petit Chien qui secoue de l'argent et des pierreries" present two consecutive acts of deception that are juxtaposed and complementary because one is a response to the other. A deceit of secondary importance may be related to the major deceit of a tale as a concomitant frame; in "La Mandragore" or "L'Ermite," for example, a lie in the first tale and a disguise in the second subtend the whole deceptive process. Lies sometimes operate as a preliminary instigating factor. In Appendix B I have summarized the various occurrences and classifications of the patterns of deceit that were described more amply in my text.

TOWARDS A POETICS OF COGNITIVE SPACE

In his conclusion to *Figures of the Text* Michael Vincent comments: "All branches of poetics could no doubt be enriched through their contact with La Fontaine" (110). I, too, believe that La Fontaine's texts present a fertile ground for developing critical theory. In this study I have developed a critical method that can be applied not only to the *Contes* of La Fontaine but to other literary texts as well. The cognitive dimension of literature is an area of research that deserves further investigation; I propose here one method for doing so, namely, frame analysis. As early as 1975 Greimas tentatively, and somewhat unobtrusively, coined the term "*dimension cognitive*" in the final pages of *La Sémiotique du texte*:

> Un fait nouveau s'est imposé progressivement à l'évidence: c'est l'existence de la *dimension cognitive* de la narrativité qui en arrive à doubler pratiquement l'ensemble du récit et dont nous avons été obligé, non sans tâtonnements et interprétations partielles, d'affirmer l'autonomie... La reconnaissance de cette nouvelle dimension dont, pour l'instant, on voit mal la configuration d'ensemble n'est qu'une ouverture vers de nouveaux champs d'explorations. (265)

A few years later he explored further the notion of the cognitive dimension by applying his theory to a study of Maupassant's short story, "La Ficelle":

> On voit d'ailleurs bien qu'à ce savoir différencié qui se trouve ainsi introduit—tel acteur sachant telle chose et tel autre une autre chose—s'oppose, dans la première partie du récit, le *savoir absolu* du sujet de la narration qui, prenant le lecteur pour complice, parle des gens et des choses comme s'il était omniprésent et omniscient. (*Du Sens II* 144)

The intuitions and preliminary probing into the cognitive dimension of literature on the part of Greimas were, unfortunately, not immediately developed by other literary theoreticians.[4] Critical theory has, however, changed so significantly since the time of the early structuralists and the New Critics, who virtually abandoned questions relating directly to the story world and particularly to the characters in the diegetic world, that I dare hope that now the time has come to develop a new poetics of cognitive space. This project, to which I have contributed with the present study of deceit in La Fontaine's *Contes*, might well be fruitfully pursued, at least initially, by delving into other seventeenth-century texts and examining, for example, the role of secrecy in the plays of Racine, irony in the theater of Molière, or the functioning of lies in the dramas of Corneille.

NOTES

1. See my article, "The Optics of Relativism in the *Fables* of La Fontaine," where I establish a taxonomy of Lafontainian relativism by analyzing a number of fables that exemplify the various categories of relativism.

2. See Greimas and Courtés, *Sémiotique: Dictionnaire raisonné*, vols. 1 and 2.

3. Two *contes*, "La Gageure des trois commères" and "Le Roi Candaule et le Maître en droit," contain more than one independent tale with only a slight connecting link.

4. The situation is changing somewhat, particularly in the area of the representation of subjectivity in literature—see especially J. Fontanille, *Les Espaces subjectifs*.

APPENDIX A

ALPHABETICAL LIST OF CONTES

APPENDIX A

ALPHABETICAL LIST OF CONTES

1. A femme avare galant escroc
2. L'Abbesse
3. L'Anneau de Hans Carvel
4. Les Aveux indiscrets
5. Belphégor
6. Le Berceau
7. Le Calendrier des vieillards
8. Le Cas de conscience
9. La Chose impossible
10. La Clochette
11. Le Cocu battu et content
12. Comment l'esprit vient aux filles
13. La Confidente sans le savoir ou le Stratagème
14. Conte d'une chose arrivée à Château-Thierry
15. La Coupe enchantée
16. La Courtisane amoureuse
17. Le Cuvier
18. Le Diable de Papefiguière
19. Le Diable en enfer
20. L'Ermite
21. Le Faiseur d'oreilles et le raccommodeur de moules
22. Le Faucon
23. Féronde ou le Purgatoire
24. La Fiancée du roi de Garbe
25. Le Fleuve Scamandre

26. Les Frères de Catalogne
27. La Gageure des trois commères
28. Le Gascon puni
29. Joconde
30. La Jument du compère Pierre
31. Les Lunettes
32. Le Magnifique
33. La Mandragore
34. Le Mari confesseur
35. La Matrone d'Ephèse
36. Mazet de Lamporechio
37. Le Muletier
38. Nicaise
39. Les Oies de frère Philippe
40. On ne s'avise jamais de tout
41. L'Oraison de saint Julien
42. Pâté d'Anguille
43. Le Petit Chien qui secoue de l'argent et des pierreries
44. Le Psautier
45. Les Quiproquos
46. Le Remède
47. Les Rémois
48. Richard Minutolo
49. Le Roi Candaule et le Maître en droit
50. La Servante justifiée
51. Le Tableau
52. Les Troqueurs

NOTE

These are the *contes* that I consider as truly belonging to the genre, as it was defined by La Fontaine. A number of other poems were published in the *Contes et nouvelles*, but they do not meet the strictest definitions of the genre. I have explained the criteria for selection of this corpus near the end of the Introduction.

APPENDIX B

CLASSIFICATION OF TALES ACCORDING TO DOMINANT INSTRUMENTS OF DECEIT AND PATTERNS OF DECEIT

APPENDIX B

CLASSIFICATION OF TALES ACCORDING TO DOMINANT INSTRUMENTS OF DECEIT AND PATTERNS OF DECEIT

1. MAGIC OR PSEUDOMAGIC

La Coupe enchantée: *Embedded Frames*
Féronde ou le Purgatoire: *Simple Deceit Frame*
La Gageure des trois commères (2nd tale): *Embedded Frames*
La Jument du compère Pierre: *Simple Deceit Frame*
Le Petit Chien qui secoue de l'argent et des pierreries: *Juxtaposed Frames*

2. DISGUISE

Le Cocu battu et content: *Embedded Frames*;
Preliminary Deceit: Lie
Le Fleuve Scamandre: *Simple Deceit Frame*
La Gageure des trois commères (1st tale): *Embedded Frames*
Les Lunettes: *Embedded Frames*
La Mandragore: *Juxtaposed Frames*;
Concomitant Deceit: Lie
Le Mari confesseur: *Embedded Frames*
Le Muletier: *Juxtaposed Frames*

3. DISPLACEMENT

Le Gascon puni: *Embedded Frames*;
Concomitant Deceit: Lie
Joconde: *Juxtaposed and Embedded Frames*
Concomitant Deceit: Replacement
Les Quiproquos: *Naive Juxtaposed Frames*
Le Remède: *Simple Deceit Frame*

Richard Minutolo: *Embedded Frames*;
Concomitant Deceit: Lie
Le Roi Candaule et le Maître en droit (2nd tale): *Embedded Frames*
La Servante justifiée: *Embedded Frames*

4. MISPLACEMENT

L'Anneau de Hans Carvel: *Naive Simple Deceit Pattern*
Le Berceau: *Naive Juxtaposed Frames*
Le Psautier: *Naive Simple Deceit Frame*

5. REPLACEMENT

La Courtisane amoureuse: *Naive Simple Deceit*
Le Faucon: *Naive Simple Deceit*
La Matrone d'Ephèse: *Naive Juxtaposed Frames*
Le Roi Candaule et le Maître en droit (1st tale): *Naive Embedded Frames*
Les Troqueurs: *Naive Embedded Frames*
Le Tableau: *Juxtaposed Frames*

6. LYING

Les Aveux indiscrets: *Juxtaposed Frames*
Belphégor: *Simple Deceit Frame*
La Chose impossible: *Simple Deceit Frame*
La Clochette: *Simple Deceit Frame*
La Confidente sans le savoir ou le Stratagème: *Embedded Frames*
Le Cuvier: *Simple Deceit Frame*
Le Diable de Papefiguière: *Simple Deceit Frame*
La Gageure des trois commères (3rd tale): *Simple Deceit Frame*
Mazet de Lamporechio: *Embedded Frames*
Les Oies de frère Philippe: *Simple Deceit Frame*
On ne s'avise jamais de tout: *Simple Deceit Frame*
Les Rémois: *Embedded Frames*
Preliminary Deceit: Lie

7. FALSE PROMISE

A femme avare galant escroc: *Embedded Anticipation Frame*
Comment l'esprit vient aux filles: *Embedded Anticipation Frame*;
Preliminary Deceit: Lie
Conte d'une chose arrivée à Château-Thierry: *Embedded Anticipation Frame*
Le Diable en enfer: *Embedded Anticipation Frame*

L'Ermite: *Embedded Anticipation Frame*;
Concomitant Deceit: Disguise
Le Faiseur d'oreilles et le raccommodeur de moules:
Juxtaposed Embedded Anticipation Frames;
Preliminary Deceit: Lie
Le Magnifique: *Juxtaposed Embedded Anticipation Frames*
Nicaise: *Naive Embedded Anticipation Frame*
L'Oraison de saint Julien: *Embedded Anticipation Frame*

8. CASUISTRY

L'Abbesse: *Simple Casuistical Frames*
Le Calendrier des vieillards: *Juxtaposed Casuistical Frames*
Le Cas de conscience: *Juxtaposed Casuistical Frames*
La Fiancée du roi de Garbe: *Simple Casuistical Frames*
Les Frères de Catalogne: *Juxtaposed Casuistical Frames*
Pâté d'Anguille: *Juxtaposed Casuistical Frames*

* See the Conclusion for an explanation about the types of deceit frames

WORKS CITED

Aarne, Antti, and Stith Thompson. *The Types of the Folktale: A Classification and Bibliography.* 2nd rev. FF Communications 184. Helsinki: Tampere, 1964.

Almansi, Guido. *The Writer as Liar: Narrative Technique in the Decameron.* London: Routledge & Kegan Paul, 1975.

Aretino, Pietro. *Ragionamenti: Sei Giornate.* Ed. Giovanni Aquilecchia. Bari: G. Laterza, 1969.

Ariosto, Ludovico. *Orlando furioso.* 2 vols. Edited by Remo Ceserani. Torino: Unione Tipographico, 1963.

Austin, J. L. "Performative Utterances." In *Philosophical Papers*, 233-52. 3rd ed. Oxford: Oxford University Press, 1979.

Banville, Théodore de. "Jean de La Fontaine." In *Petit traité de versification française*, 301-28. Paris: Fasquelle, 1899.

Barthes, Roland. *S/Z.* Coll. "Tel Quel," Paris: Seuil, 1970.

Bates, Robert, and Gregory C. Richter, "The Capon Theory of the Cuckold's Horns: Confirmation or Conjecture?" *Journal of American Folklore* 100 (1987): 58-63.

Bateson, Gregory. *Steps to an Ecology of Mind.* New York: Ballantine, 1972.

Bateson, Gregory, and Jurgen Ruesch. *Communication: The Social Matrix of Psychiatry.* New York: Norton, 1968.

Bauny, Estienne. *La Somme des péchés.* Caen: Corneille Pitreson, 1653.

Benveniste, *Problèmes de linguistique générale.* Paris: Gallimard, 1966.

Beugnot, Bernard, and Roger Zuber. *Boileau, visages anciens, visages modernes.* Montréal: Presses Universitaires de Montréal, 1973.

Birberick, Anne L. "From World to Text: The Figure of the Nun in La Fontaine's *Contes*." In *Refiguring La Fontaine: Tercentenary Essays*, edited by Anne L. Birberick, 181-201. Charlottesville: Rookwood Press, 1996.

Boccaccio. *Il Decamerone*. Ed. G. Macchia. Milan: Adelphi, 1982.

—. *The Decameron*. Ed. Jonathan Usher. Trans. Guido Waldman. The World's Classics. Oxford: Oxford University Press, 1993.

Boileau, Nicolas. "Dissertation sur Joconde." In *Oeuvres complètes*, edited by Françoise Escal and Antoine Adam, 309-24. Bibliothèque de la Pléiade. Paris: Gallimard, 1966.

Bok, Sissela. *Lying: Moral Choice in Public and Private Life*. New York: Pantheon, 1978.

Booth, Wayne. *The Rhetoric of Fiction*. Chicago: University of Chicago Press, 1961.

Börnas, Göran. "Le Cocu battu et content: Etude sur un conte de La Fontaine." *Studia Neophilologica* 44 (1972): 37-61.

Brusegan, Rosanna. "Les Fonctions de la ruse dans les fabliaux." *Strumenti critici: Rivista Quadrimestrale* 16 (1982): 148-60.

Bruster, Douglas. "The Horn of Plenty: Cuckoldry and Capital in the Drama of the Age of Shakespeare." *Studies in English Literature, 1500-1900* 30 (1990): 195-215.

Bryan, W. F., and Germaine Dempster. *Sources and Analogues of Chaucer's Canterbury Tales*. New York: Humanities Press, 1941.

Callières, François de, ed. *Des bons mots et des bons contes*. Paris: Barbin, 1692.

Cariou, Pierre. *Pascal et la casuïstique*. Paris: PUF, 1993.

Cauley, Joseph. "Narrative Technique in the *Contes et nouvelles en vers* of La Fontaine." *French Literature Series* 2 (1975): 27-38.

Chambers, Ross. *Story and Situation: Narrative Seduction and the Power of Fiction*. Minneapolis: University of Minnesota Press, 1984.

Chandernagor, Françoise. *L'Allée du roi: Souvenirs de Françoise d'Aubigné, marquise de Maintenon, épouse du roi de France*. Paris: Julliard, 1981.

Chatman, Seymour. *Coming to Terms: The Rhetoric of Narration in Fiction and Film*. Ithaca and London: Cornell University Press, 1990.

—. *Story and Discourse: Narrative Structure in Fiction and Film*. Ithaca: Cornell University Press, 1978.

Chaucer, Geoffrey. *The Complete Works of Geoffrey Chaucer*. Ed. Walter W. Skeat. Vol. 4. *The Canterbury Tales*. Oxford: Clarendon Press, 1900.

Works Cited

Cheeney, Dorothy L., and Robert M. Seyfarth. *How Monkeys See the World.* Chicago and London: University of Chicago Press, 1990.

Cioranescu, Alexandre. *L'Arioste en France des origines à la fin du XVIIIe siècle.* 2 vols. Paris: Editions des Presses modernes, 1939.

Clarac, Pierre. *La Fontaine par lui-même.* Paris: Seuil, 1961.

—. "La Fontaine." *Littérature française.* Vol. I. Ed. Antoine Adam et al. Paris: Larousse, 1967.

Collinet, Jean-Pierre. "L'Amour et son expression poétique dans les *Contes* de La Fontaine." *Travaux de linguistique et de littérature* 18, no. 2 (1980): 109-14.

—. "La Fontaine est-il poète?" *Oeuvres et critiques* 5 (1980): 259-91.

—. "La Matière et l'art du prologue dans les *Contes* de La Fontaine." *Studi francesi* 74 (1981): 219-37.

Combe, Dominique. *Les Genres littéraires.* Paris: Hachette, 1992.

Conteurs français du XVIe siècle. Ed. Pierre Jourda. Bibliothèque de la Pléiade. Paris: Gallimard, 1979.

Culler, Jonathan. *Structuralist Poetics: Structuralism, Linguistics, and the Study of Literature.* Ithaca: Cornell University Press, 1976.

Dandrey, Patrick. *La Fabrique des Fables: Essai sur la poétique de La Fontaine.* Paris: Klincksieck, 1992.

Danner, Richard. *Patterns of Irony in the Fables of La Fontaine.* Athens: Ohio University Press, 1985.

Defrenne, Madeleine. "La Fontaine et les jeux de focalisation." *Cahiers de Littérature du XVIIème siècle* 6 (1984): 321-30.

De Waal, F. "Deception in the Natural Communication of Chimpanzees." In *Deception: Perspectives on Human and Nonhuman Deceit,* edited by R.W. Mitchell and N.S. Thomson, 22-44. Albany: State University of New York Press, 1986.

Dinsmore, John. "Mental Spaces from a Functional Perspective." *Cognitive Science* 11 (1987): 1-21.

Dubuis, Roger. "La Fontaine lecteur des *Cent Nouvelles nouvelles.*" In *Mélanges de littérature et d'histoire offerts à Georges Couton,* 345-57. Lyon: Presses Universitaires, 1981.

Ducrot, Oswald, and Tzvetan Todorov. *Dictionnaire encyclopédique des sciences du langage.* Paris: Seuil, 1972.

Durand, Gilbert. *Les Structures anthropologiques de l'imaginaire.* Paris: PUF, 1960.

Eck, Marcel. *Lies and Truth*. Trans. B. Murchland. New York: MacMillan, 1970.

Eco, Umberto. *Opera aperta*. Milan: Bompiani, 1962.

Ejxenbaum, Boris M. "The Theory of the Formal Method." In *Readings in Russian Poetics: Formalist and Structuralist Views*, edited by Ladislav Matejka and Krystyna Pomorska, 3-37. Boston: MIT Press, 1971.

Escobar, Anthonius. *Liber theologiae moralis*. New ed. Lugduni, 1646.

Ezquerro, Milagros. "Les connections du système PERSE." In *Actes du Colloque de Toulouse : Le Personnage en question*, 103-9. Toulouse: Toulouse-Le Mirail, 1984.

Fauconnier, Gilles. *Mental Spaces: Aspects of Meaning Construction in Natural Language*. Cambridge: Harvard University Press, 1974.

Felman, Shoshana. *The Literary Speech Act: Don Juan with J.L. Austin, or Seduction in Two Languages*. Cornell University Press, 1983.

Ferreyrolles, Gérard. *Blaise Pascal: Les Provinciales*. Paris: PUF, 1984.

Fillmore, Charles J. "An Alternative to Checklist Theories of Meaning." In *Proceedings of the First Annual Meeting of the Berkeley Linguistics Society*, 123-32. Berkeley: Linguistics Department, 1975.

Fingarette, Herbert. *Self-Deception*. London: Routledge & Kegan Paul, 1969.

Fontanille, Jacques. *Le Savoir partagé: Sémiotique et théorie de la connaissance chez Marcel Proust*. Paris: Hadès, 1987,

—. *Les Espaces subjectifs: Introduction à la sémiotique de l'observateur*. Paris: Hachette, 1989.

Forestier, Georges. *Esthétique de l'identité dans le théâtre français (1550-1680): le Déguisement et ses avatars*. Geneva: Droz, 1988.

Frake, Charles O. "Plying Frames Can Be Dangerous: Some Reflections on Methodology in Cognitive Anthropology." *The Quarterly Newsletter of the Institute for Comparative Human Cognition* 1 (1977): 1-7.

Franco, Charles. "Boccaccio's Woman: A Liberated Woman?" *Mid-Hudson Language Studies* 4 (1981): 13-23.

Furetière, Antoine. *Dictionnaire universel*. 3 vols. 1690. Rpt. Geneva: Slatkine, 1970.

Gass, W. H. *Fiction and the Figures of Life*. Boston: Godine, 1979.

Genette, Gérard. *Figures III*. Paris: Seuil, 1972.

Genot, Gérard. "La Fontaine et Boccace." In *Il Boccaccio nella cultura francese*, edited by Carlo Pellegrini, 567-79. Florence, Leo S. Olschki, 1971.

—. "Le récit [du] déclassé." *Revue Romaine* 7 (1972): 204-32.

Goffman, Erving. *Frame Analysis: An Essay on the Organization of Experience*. Cambridge: MIT Press, 1974.

Gregorio, Laurence A. *The Pastoral Masquerade: Disguise and Identity in "L'Astrée."* Saratoga, Cal.: Anma Libri, 1992.

Greimas, A.-J. "The Cognitive Dimension of Narrative Discourse." *New Literary History* 7 (1975-76): 433-47.

—. *Du sens II*. Paris: Seuil, 1983.

—. *Maupassant: La Sémiotique du texte*. Paris: Seuil, 1975.

—. "Un Problème de sémiotique narrative: Les Objets de valeur." *Langages* 31 (1973): 13-35.

Greimas, A.-J., and J. Courtés. *Sémiotique: Dictionnaire raisonné de la théorie du langage*. 2 vols. Paris: Hachette, 1979, 1986.

Grimm, Jürgen. "'Le Faucon.' La Fontaine et Boccace: Proximité et distance." In *Correspondances: Mélanges offerts à Roger Duchêne*, edited by Wolfgang Leiner and Pierre Ronzeaud, 215-24. Tübingen: Gunter Narr Verlag, 1992.

Grisé, Catherine. "La Casuistique dans *Les Contes* de La Fontaine." *Studi francesi* 99 (1990): 411-21.

—. "The *Conte en vers*: Expanding Stith Thompson's X-File of Obscene Motifs." *Folklore* 108 (1997): 25-34.

—. "The Discourse of Seductive Promises in the *Contes* of La Fontaine." *Cincinnati Romance Review* 11 (1992): 53-59.

—. "Disguise in the *Contes* of La Fontaine: The Manipulation of Cognitive Space." *Cahiers du Dix-septième* 4 (1990): 411-21.

—. "Erotic Dimensions of Space in La Fontaine's `La Fiancée du roi de Garbe.'" *Modern Language Review* 82 (1987): 587-98.

—. "Ironie dramatique ou ironie cognitive?" *Neophilologus* 74 (1990): 353-60.

—. "Le Jeu de l'imitation: Un Aspect de la réception des *Contes* de La Fontaine." *PFSCL* 18 (1983): 249-62.

—. "The Optics of Relativism in the *Fables* of La Fontaine." In *Refiguring La Fontaine: Tercentenary Essays*, edited by Anne L. Birberick, 126-56. Charlottesville: Rookwood Press, 1996.

—. "Under the Pear Tree: Cognitive Space and Deceit Structures in Five 'Magical' *Contes* of La Fontaine." *French Forum* 16 (1991): 21-38.

Gsteiger, Manfred. "'La Fiancée du roi de Garbe': De Boccace à La Fontaine". In *Sensus communis: Contemporary Trends in Comparative Literature*, edited by Janos Riesz, Peter Boerner, Bernard Scholz, 325-33. Tubingen: Narr, 1986.

Guichard, Jean-François. *Contes et autres poésies*. Paris: Suret, 1802.

Guiraud, Pierre. *Dictionnaire érotique*. 3rd ed. Paris: Payot, 1993.

—. *Etymologie des mots obscènes*. Paris: Payot, 1982.

Hahn, Thomas. "Money, Sexuality, Wordplay, and Context in *The Shipman's Tale*." In *Chaucer in the Eighties*, edited by Julian N. Wasserman and Robert J. Blanch, 235-49. Syracuse: Syracuse University Press, 1986.

Haight, M. R. *A Study of Self-Deception*. Sussex: Harvester Press, 1980.

Hamon, Philippe. "Pour un statut sémiologique du personnage." *Littérature* 6 (1972): 86-110. Reedited in *Poétique du récit* by R. Barthes, W. Kayser, W. Booth, Ph. Hamon, 115-80. Paris: Seuil, 1977.

Hochman, Baruch. *Character in Literature*. London: Cornell University Press, 1985.

Houle, Martha M. "La Fontaine's Fun with Casuistry in `Le Cas de conscience'." *Papers on French Seventeenth-Century Literature* 14 (1987): 193-215.

—. "The Play of *bienséances* in La Fontaine's 'L'Abbesse.'" *Papers on French Seventeenth-Century Literature* 18 (1991): 109-21.

Howard, Fannie Scott. "La Fontaine on Fiction Writing: Reality and Illusion in the *Contes*." *French Literature Series* 2 (1975): 167-71.

Hunter, Dianne. *Seduction and Theory*. Chicago: University of Illinois Press, 1989.

Hymes, Dell. "Ways of Speaking." In *Explorations in the Ethnography of Speaking*, edited by Richard Bauman and Joel Sherzer, 433-51. Cambridge: Cambridge University Press, 1974.

Jakobson, Roman. "Qu'est-ce que la poésie?" Trans. Marguerite Derrida. In *Questions de poétique*, 113-26. Paris: Seuil, 1973.

Johnson, Lesley. "Women on Top: Antifeminism in the fabliaux?" *MLR* 78 (1983): 298- 307.

Jonsen, Albert R., and Stephen Toulmin. *The Abuse of Casuistry*. Berkeley: University of California Press, 1988.

Jost, François. *L'Oeil-caméra, entre film et roman*. Lyon: Presses Universitaires de Lyon, 1987.

Kerbrat-Orecchionni, C. *L'Enonciation: De la subjectivité dans le langage*. Paris: Armand Colin, 1980.

Krueger, Roberta L. "Love, Honor, and the Exchange of Women in *Yvain*: Some Remarks on the Female Reader." *Romance Notes* 25 (1985): 302-17.

Kuizenga, Donna. "La Fontaine's Le 'Faucon': A Lesson of Experience." *French Forum* 2 (1977): 214-23.

Lacey, Norris J. "Fabliaux Women." *Romance Notes* 25 (1985): 318-27.

La Fontaine, Jean de. *Contes et nouvelles en vers*. Ed. Nicole Ferrier and Jean-Pierre Collinet. Paris: Garnier-Flammarion, 1980.

—. *Contes et nouvelles en vers*. Ed. Alain-Marie Bassy. Paris: Gallimard, 1982.

—. *Oeuvres complètes* I. Ed. Jean-Pierre Collinet. Bibliothèque de la Pléiade. Paris: Gallimard, 1991.

—. *Oeuvres de J. de La Fontaine*. Ed. Henri Régnier. 12 vols. Les Grands Ecrivains de la France. Paris: Hachette, 1883-92.

—. *Oeuvres diverses*. Ed. Pierre Clarac. Bibliothèque de La Pléiade. Paris: Gallimard, 1958.

Lajarte, Philippe de. "Du conte facétieux considéré comme genre: Esquisse d'une analyse structurale." *Ethnologie française* 4 (1974): 319-32.

Lapp, John C. *The Esthetics of Negligence: La Fontaine's "Contes."* Cambridge: Cambridge University Press, 1971.

—. "The Esthetics of Negligence: La Fontaine's *Contes*." *L'Esprit créateur* 3 (1963): 108-15.

Legman, Gershon. "Towards a Motif-Index of Erotic Humor." *Journal of American Folklore* 75 (1962): 227-48.

Le Maçon, Antoine-Jean. *Le Décaméron de Jean Boccace*. Nice: Le Chant des sphères, 1976-78.

Lloyd, J. E. "Firefly Communication and Deception: 'Oh, what a tangled web.'" In *Deception: Perspectives on Human and Nonhuman Deceit*, edited by R. W. Mitchell and N.S. Thomson, 113-28. Albany: State University of New York Press, 1986.

Lorcin, Marie-Thérèse. *Façons de sentir et de penser: Les Fabliaux français*. Paris: Champion, 1979.

—. "Le Mariage dans les fabliaux français." In *Epopée animale, fable, fabliau*. Actes du IVe Colloque de la Société internationale renardienne, edited by Gabriel Bianciotto and Michel Salvat, 333-43. Paris: PUF, 1984.

Lugli, Vittorio. "La Courtisane amoureuse." In *Studi in onore di Italo Siciliano* 2: 711-20. Florence: Leo S. Olschki, 1966.

Lyons, John D. "Author and Reader in the *Fables*." *French Review* 49 (1975): 59-67.

—. "D'une vérité sans effet: La Fontaine et la théorie de la nouvelle." *Littératures classiques* 11 (1989): 109-20.

Manning, Peter K., and Keith Hawkins. "Legal Decisions: A Frame Analytic Perspective." In *Beyond Goffman*, edited by Stephen Harold Riggins, 203-33. New York: Mouton de Gruyter, 1990.

Maus, Katharine Eisaman. "Horns of Dilemma: Jealousy, Gender, and Spectatorship in English Renaissance Drama." *English Literary History* 54 (1987): 561-83.

McLaughlin, Brian P. "Exploring the Possibility of Self-Deception in Belief." In *Perspectives on Self-Deception*, edited by Brian P. McLaughlin and Amélie Oksenberg Rorty, 29-62. Berkeley: University of California Press, 1988.

McLaughlin, Brian P., and Amélie Oksenberg Rorty, eds. *Perspectives on Self-Deception*. Berkeley: University of California Press, 1988.

Merino-Morais, Jane. *Différence et répétition dans les "Contes" de La Fontaine*. University of Florida Monographs: Humanities 52. Gainesville: University Press of Florida, 1981.

Mikhailov, A. D. "Genèse de la satire antibourgeoise dans les fabliaux." In *Epopée animale, fable, fabliau*. Actes du IVe Colloque de la Société internationale renardienne, edited by Gabriel Bianciotto and Michel Salvat, 355-63. Paris: PUF, 1984.

Miles, H. L. "How Can I Tell a Lie? Apes, Language and the Problem of Deception." In *Deception: Perspectives on Human and Nonhuman Deceit*, edited by R. W. Mitchell and N. S. Thomson, 245-66. Albany: State University of New York Press, 1986.

Mitchell, R. W., and N. S. Thomson, eds. *Deception: Perspectives on Human and Nonhuman Deceit*. Albany: State University of New York Press, 1986.

Mongrédien, Georges. *Recueil des textes et des documents du XVIIe siècle relatifs à La Fontaine*. Paris: CNRS, 1973.

Montagut, Emile. "'La Fiancée du roi de Garbe' et le *Décaméron*." *Revue des deux mondes*, 45 (1863): 721-36.

Montaiglon, A. de, and G. Raynaud, eds. *Recueil général et complet des fabliaux des XIIIe et XIVe siècles*. 6 vols. Paris, 1872-90. Slatkine Reprints, 1973, 3 vols.

Works Cited

Muscatine, Charles. *The Old French Fabliaux*. New Haven: Yale University Press, 1986.

Nisard, D. *Histoire de la littérature française*. Vol. 3. Paris: Firmin Didot, 1863.

Nurmela, Tauno. "La Misogynie chez Boccace." In *Boccaccio in Europe*. Proceedings of the Boccaccio Conference, Louvain 1975, edited by Gilbert Tournoy, 191-96. Louvain: Presses Universitaires Louvain, 1977.

Nuyts, Jan. "Epistemic Modal Verbs and Adjectives and the Layered Representation of Conceptual and Linguistic Structure." *Linguistics* 31 (1993), 933-69.

Olsen, Michel. *Les Transformations du triangle érotique*. Copenhagen: Akademisk Forlag, 1976.

Orieux, Jean. *La Fontaine*. Paris: Flammarion, 1976.

Ovid. *"The Art of Love," and Other Poems. Ovid in Six Volumes*. Vol. 2. Trans. J. H. Mozley. Loeb Library. London: William Heinemann, 1969.

Paris, Gaston. "La Source italienne de `La Courtisane amoureuse.'" In *Raccolta di studi critice dedicati ad A. d'Ancona*, 375-85. Florence: G. Barbera, 1901.

Pascal, Blaise. *Oeuvres complètes: Les Provinciales*. Ed. Jacques Chevalier. Bibliothèque de la Pléiade. Paris: Gallimard, 1954.

Pavel, Thomas G. *Fictional Worlds*. Cambridge: Harvard University Press, 1986.

Picard, Michel. *La Lecture comme jeu*. Paris: Éditions du Minuit, 1986.

Plantié, Jacqueline. "L'Aboutissement de la tradition facétieuse dans une fable et un conte de La Fontaine." *Revue d'histoire littéraire de la France* 84 (1984): 531-38.

Pollard-Gott, L. "Attribution Theory and the Novel." *Poetics* 21 (1993): 499-524.

Poughon, Jean-Michel. *Histoire doctrinale de l'échange*. Paris: R. Pichon and R. Durand-Auzias, 1987.

Prince, Gerald. *Dictionary of Narratology*. Lincoln and London: University of Nebraska Press, 1987.

Propp, Vladimir. *Morphologie du conte*. Paris: Seuil, 1965.

Ricoeur, Paul. *La Narration: Quand le récit devient communication*. Paris: Labor et Fides, 1988.

Rimmon-Kenan, Shlomith. *Narrative Fiction: Contemporary Poetics.* London and New York: Methuen, 1983.

Rorty, Amélie Oksenberg. "The Deceptive Self: Liars, Layers, and Lairs." In *Perspectives on Self-Deception*, edited by Brian P. McLaughlin and Amélie Oksenberg Rorty, 11-28. Berkeley: University of California Press, 1988.

Rubin, David Lee. *A Pact with Silence: Art and Thought in the Fables of Jean de La Fontaine.* Columbus: Ohio State University Press, 1991.

Rubin, Gayle. "The Traffic in Women: Notes on the 'Political Economy' of Sex." In *Toward an Anthropology of Women*, edited by Rayna R. Reiter, 157-210. New York and London: Monthly Review Press, 1975.

Saint-Glas, Pierre de. *Contes nouveaux en vers.* Paris: P. Tribouillet, 1672.

Saint-Mard, Rémond de. *Réflexions sur la poésie en général.* Paris: Delespine, 1734.

Scheibe, Karl E. *Mirrors, Masks, Lies and Secrets: The Limits of Human Predictability.* New York: Praeger Publishers, 1979.

Schenck, Mary Jane Stearns. *The Fabliaux: Tales of Wit and Deception.* Purdue University Monographs in Romance Languages. Vol. 24. Amsterdam and Philadelphia: John Benjamins, 1987.

Shell, Marc. *The Economy of Literature.* Baltimore and London: Johns Hopkins University Press: 1978.

Smith, Barbara Herrnstein. *Poetic Closure: A Study of How Poems End.* Chicago: University of Chicago Press, 1968.

Soriano, Marc. "Des *Contes* aux *Fables*." *Europe* 515 (1972): 99-131.

Sweetser, Marie-Odile. "Images féminines chez La Fontaine." In *Correspondances: Mélanges offerts à Roger Duchêne*, edited by Wolfgang Leiner and Pierre Ronzeaud, 201-13. Tübingen: Gunter Narr Verlag, 1992.

—. *La Fontaine.* Twayne's World Authors Series. Boston: Twayne Publishers, 1987.

—. "Une Fugue à trois voix: 'Joconde.'" *Littératures classiques* 12 (1990): 213-24.

Tannen, Deborah, ed. *Framing in Discourse.* New York: Oxford University Press: 1993.

—. "What's in a Frame? Surface Evidence for Underlying Expectations." In *Framing in Discourse*, edited by Deborah Tannen, 14-56. New York: Oxford University Press: 1993.

Thomas, Keith. "The Double Standard." *Journal of the History of Ideas* 20 (1959): 195- 216.

Thompson, Stith. *Motif-Index of Folk Literature*. 6 vols. Copenhagen: Rosenkilde and Bagger, 1955-1958.

Todorov, Tzvetan. *Grammaire du Décaméron*. The Hague: Mouton, 1969.

Tyler, J. Allen, ed. *A Concordance to the "Fables" and "Contes" of Jean de La Fontaine*. Ithaca: Cornell University Press, 1974.

Vaillant, Alain. *La Poésie*. Paris: Nathan, 1992.

Vareecke, Louis-Gustave. "Casuïstique." In *Encyclopaedia universalis* 4: 347-48. Paris, 1984.

Vaughan, Geneviève "Communication and Exchange." *Semiotica* 29 (1980): 113-43.

Vincent, Michael. "Reading (through) the Veil: 'Le Tableau'." In *Figures of the Text: Reading and Writing (in) La Fontaine*, 91-108. Amsterdam and Philadelphia: John Benjamins, 1992.

Walton, K. "How Remote are Fictional Worlds from the Real World?" *Journal of Aesthetics and Art Criticism* 37 (1978): 11-23.

Wilson, R. "The Bright Chimera: Character as a Literary Term." *Critical Inquiry* 5 (1979): 725-49.

Wittgenstein, Ludwig. *Philosophical Investigations*. Trans. G. E. M. Anscombe. New York: Macmillan, 1953.

Zuckerman, Miron, B. M. De Paolo, and R. Rosenthal. "Verbal and Non-Verbal Communication of Deception." In *Advances in Experimental Social Psychology*. Vol. 14, edited by L. Berkowitz, 1-59. New York: Academic Press, 1981.

INDEX